SOTHIS PRESS

Sothis Press aims to completely reimagine the public domain book with an emphasis on universal access to captivatingly designed print editions. We believe that technology will enable every single book ever published to be perpetually accessible in beautiful print form. Sothis is beauty in ubiquity.

www.sothis.press
1 Waterhouse Square, London, ECIN 2ST

A Text-Book of Precious Stones

For Jewelers and the Gem-Loving Public

Frank B. Wade

SOTHIS PRESS

ISBN 978-1-0368-0291-2

A Text-Book of Precious Stones
is in the public domain.

This edition is published by
Sothis Press © 2023. The company
is registered in England and Wales.
The Company Number is 14900389.

A TEXT-BOOK
OF
PRECIOUS STONES

FOR JEWELERS
and
THE GEM-LOVING PUBLIC

BY

FRANK B. WADE, B.S.

HEAD OF THE DEPARTMENT OF CHEMISTRY, SHORTRIDGE HIGH SCHOOL, INDIANAPOLIS, IND.

AUTHOR OF "DIAMONDS: A STUDY OF THE FACTORS THAT GOVERN THEIR VALUE"

ILLUSTRATED

G. P. PUTNAM'S SONS
NEW YORK AND LONDON
The Knickerbocker Press
1918

The Knickerbocker Press, New York

PREFACE

IN this little text-book the author has tried to combine the trade information which he has gained in his avocation, the study of precious stones, with the scientific knowledge bearing thereon, which his vocation, the teaching of chemistry, has compelled him to master.

In planning and in writing the book, every effort has been made to teach the fundamental principles and methods in use for identifying precious stones, in as natural an order as possible. This has been done in the belief that the necessary information will thus be much more readily acquired by the busy gem merchant or jeweler than would have been the case had the material been arranged in the usual systematic order. The latter is of advantage for quick reference after the fundamentals of the subject have been mastered. It is hoped, however, that the method of presentation used in this book will make easy the acquisition of a knowledge of gemology and that many who have been deterred from studying the subject by a feeling that the difficulties due to their lack of scientific training were insurmountable, will find that they can learn all the science that is really necessary, as they proceed. To that end the discussions have been given in as untechnical language as possible and homely illustrations have in many cases been provided.

Nearly every portion of the subject that a gem merchant needs to know has been considered and there is provided for the interested public much material which will enable them to be more intelligent purchasers of gem-set jewelry, as well as more appreciative lovers of Nature's wonderful mineral masterpieces.

F. B. W.

INDIANAPOLIS,
December 26, 1916

INTRODUCTION

BECAUSE of the rapid increase in knowledge about precious stones on the part of the buying public, it has become necessary for the gem merchant and his clerks and salesmen to know at least as much about the subject of gemology as their better informed customers are likely to know.

In many recent articles in trade papers, attention has been called to this need, and to the provision which Columbia University has made for a course in the study of gems. The action of the National Association of Goldsmiths of Great Britain in providing annual examinations in gemology, and in granting certificates and diplomas to those who successfully pass the examinations, has also been reported, and it has been suggested that some such course should be pursued by jewelers' associations in this country. The greatest difficulty in the way of such formal study among our jewelers and gem merchants is the lack of time for attendance on formal courses, which must necessarily be given at definite times and in definite places.

As a diamond salesman was heard to say recently: "The boss said he wanted me to take in that course at Columbia, but he didn't tell me how I was going to do it. Here I am a thousand miles from Columbia, and it was only six weeks ago that he was telling me I ought to take that course. I can't stay around New York all the time." Similarly those whose work keeps them in New York might object that their hours of employment prevented attendance on day courses, and that distance from the university and fatigue prevent attendance on night courses. The great mass of gem dealers in other cities must also be considered.

It will therefore be the endeavor of this book to provide guidance for those who really want to make themselves more efficient in the gem business, but who have felt that they needed something in the way of suggestion regarding what to attempt, and how to go about it.

Study of the sort that will be suggested can be pursued in spare moments, on street cars or elevated trains, in waiting rooms, or in one's room at night. It will astonish many to find how much can be accomplished by consistently utilizing spare moments. Booker T. Washington is said to have written in such spare time practically all that he has published.

For the practical study of the gems themselves, which is an absolutely essential part of the work, those actually engaged in the trade have better opportunities than any school could give and, except during rush seasons, there is plenty of time during business hours for such study. No intelligent employer will begrudge such use of time for which he is paying, if the thing be done in reason and with a serious view to improvement. The frequent application of what is acquired, as opportunity offers, in connection with ordinary salesmanship, will help fix the subject and at the same time increase sales.

Many gem dealers have been deterred from beginning a study of gems because of the seeming difficulties in connection with the scientific determination of the different varieties of stones. Now science is nothing but boiled-down common sense, and a bold front will soon convince one that most of the difficulties are more apparent than real. Such minor difficulties as exist will be approached in such a manner that a little effort will overcome them. For those who are willing to do more work, this book will suggest definite portions of particular books, which are easily available, for reference reading and study—but the lessons themselves will attempt to teach the essential things in as simple a manner as is possible.

Perhaps the first essential for the gem merchant is to be able surely to distinguish the various stones from one another and from synthetic and imitation stones.

That such ability is much needed will be clear to anyone who in casting a backward glance over his experience recalls the many serious mistakes that have come to his knowledge. Many more have doubtless occurred without detection. Several times recently the author has come across cases where large dealers have been mistaken in their determination of colored stones, particularly emeralds. Only the other day a ring was brought to me that had been bought for a genuine emerald ring after the buyer had taken it to one of the dealers in his city and had paid for an examination of it, which had resulted in its being declared genuine. On examining the stone with a lens of only moderate power, several round air bubbles were noted in it, and on barely touching it with a file it was easily scratched.

The material was green glass. Now, what was said about the dealer who sold it and the one who appraised it may be imagined. The long chain of adverse influence which will be put in action against those dealers, even though the one who sold the stone makes good the loss, is something that can be ill afforded by any dealer, and all this might have been avoided by even a rudimentary knowledge of the means of distinguishing precious stones. The dealer was doubtless honest, but, through carelessness or ignorance, was himself deceived.

Our first few lessons will therefore be concerned chiefly with learning the best means of telling the different stones from one another.

CONTENTS

A Text Book of Precious Stones

LESSON I

HOW STONES ARE DISTINGUISHED FROM ONE ANOTHER

Precious Stones Distinguished by their *Properties*. One precious stone is best distinguished from another just as substances of other types are distinguished, that is to say, by their *properties*. For example, salt and sugar are both *white*, both are *soluble in water*, and both are *odorless*. So far the italicized properties would not serve to distinguish the two substances. But sugar is *sweet* while salt is *salty* in taste. Here we have a distinguishing property. Now, just as salt and sugar have properties, so have all *precious stones*, and while, as was the case with salt and sugar, many precious stones have properties in common, yet each has also some properties which are distinctive, and which can be relied upon as differentiating the particular stone from other stones. In selecting properties for use in distinguishing precious stones, such properties as can be determined by quantity, and set down in numbers, are probably more trustworthy than those that can be observed by mere inspection. Those also which have to do with the behavior of light in passing through the stone are extremely valuable.

Importance of Numerical Properties. It is because gem dealers so often rely upon the more obvious sort of property, such as color, that they so frequently make mistakes. There may be several different types of stones of a given color, but each will be found to have its own numerical properties such as density, hardness, refractive power, dispersive power, etc., and it is only by an accurate determination of two or three of these that one can be sure what stone he has in hand. It must next be our task to find exactly what is meant by each of these numerical properties, and how one may determine each with ease and exactness.

LESSON II

REFRACTION

Explanation of Refraction. Perhaps the surest single method of distinguishing precious stones is to find out the *refractive index* of the material. To one not acquainted with the science of physics this calls for some explanation. The term *refraction* is used to describe the bending which light undergoes when it

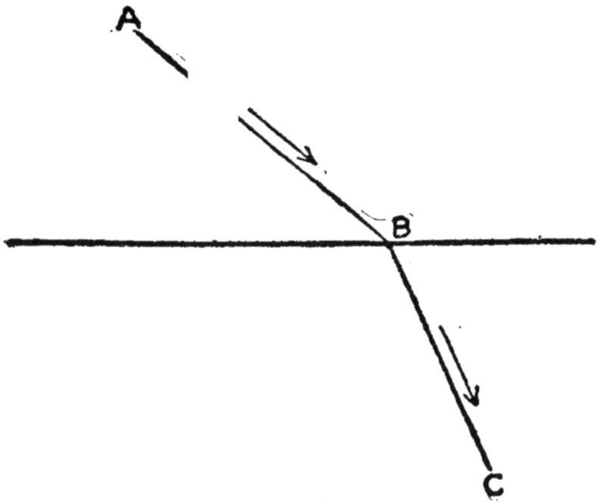

FIG. 1.

passes (at any angle but a right angle) from one transparent medium to another. For example, when light passes from air into water, its path is bent at the surface of the water and it takes a new direction within the water. (See Fig. 1.)

A B represents the path of light in the air and B C its path in the water.

While every gem stone refracts light which enters it from the air, *each stone has its own definite ability to do this, and each differs from every other in the amount of bending which it can bring about under given conditions.* The accurate determination of the amount of bending in a given case requires very finely constructed optical instruments and also a knowledge of how to apply a certain amount of mathematics. However, all this part of the work has already been done by competent scientists, and tables have been prepared by them, in which the values for each material are put down.

The Herbert-Smith Refractometer. There is

on the market an instrument called the Herbert-Smith refractometer, by means of which anyone with a little practice can read at once on the scale within the instrument the *refractive index*, as it is called, of any precious stone that is not too highly refractive. (Its upper limit is 1.80. This would exclude very few stones of importance, *i. e.*, zircon, diamond, sphene, and demantoid garnet.)

Those readers who wish to make a more intensive study of the construction and use of the refractometer will find a very full and complete account of the subject in *Gem-Stones and their Distinctive Characters*, by G. F. Herbert-Smith, New York; James Pott & Co., 1912. Chapter IV., pp. 21–36. The Herbert-Smith refractometer is there described fully, its principle is explained and directions for using it are given. The price of the refractometer is necessarily so high (duty included) that its purchase might not be justified in the case of the smaller retailer. Every large dealer in colored stones, whether importer, wholesaler, or retailer, should have one, as by its use very rapid and very accurate determinations of stones may be made, and its use is not confined to unmounted stones, for any stone whose table facet can be applied to the surface of the lens in the instrument can be determined.

LESSON III

DOUBLE REFRACTION

Explanation of Double Refraction. In Lesson II. we learned what is meant by *refraction* of light. While glass and a small number of precious stones (diamond, garnet, and spinel) bend light as was illustrated in Fig. 1, practically all the other stones cause a beam of light on entering them to separate, and the path of the light in the stone becomes double, as shown in Fig. 2.

This behavior is called *double refraction.* It may be used to distinguish those stones which are doubly refracting from those which are not. For example, in the case of a stone which is doubly refracting to a strong degree, such as a peridot (the lighter yellowish green chrysolite is the same material and behaves similarly toward light), the separation of the light is so marked that the edges of the rear facets, as seen through the table, appear ***double*** when viewed through a lens. A zircon will also similarly separate light and its rear facets also appear double lined as seen with a lens from the table of the stone. The rarer stones,

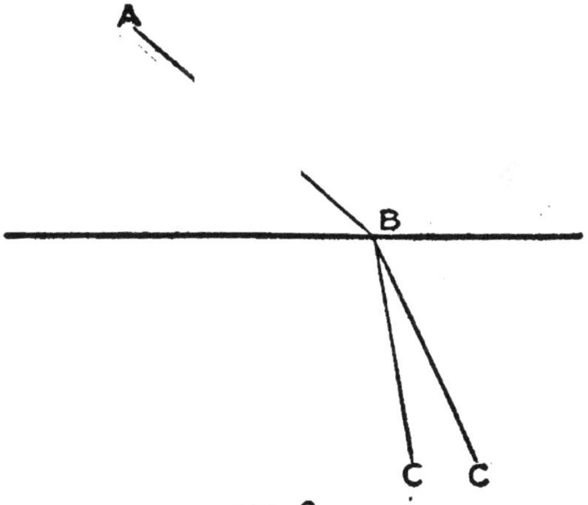

FIG. 2.

sphene and epidote, likewise exhibit this property markedly. Some colorless zircons, when well cut, so closely, resemble diamonds that even an expert might be deceived, if caught off his guard, but this simple test of looking for the doubled lines at the back of the stone would alone serve to distinguish the two stones.

A Simple but very **Valuable Test for the Kind of Refraction of a Cut Stone.** In the case of most of the other doubly refracting stones the degree of separation is much less than in peridot and zircon, and it takes a well-trained and careful eye to detect the doubling of the lines. Here a very simple device will serve to assist the eye in determining whether a cut stone is singly or doubly refracting. Expose the stone to direct sunlight and hold an opaque white card a few inches from the stone, in the direction of the sun, so as to get the bright reflections *from within the stone* reflected onto the card.

If the material is singly refractive (as in the case of diamond, garnet, spinel, and glass), *single images* of each of the reflecting facets will appear on the card, but if doubly refracting—even if slightly so—*double images* will appear. When the stone is slightly moved, these pairs of reflections will travel *together as pairs* and not tend to separate. The space between the two members of each pair of reflections serves to give a rough idea of the degree of the double refraction of the material if compared with the space between members in the case of some other kind of stone held at the same distance from the card. Thus zircon separates the reflections widely. Aquamarine, which is feebly doubly refracting, separates them but slightly.

It will be seen at once that we have here a very easily applied test and one that requires no costly apparatus. It is, furthermore, a sure test, after a little practice. For example, if one has something that looks like a fine emerald, but that may be glass, all one need to do is to expose it in the sun, as above indicated. If real emerald, double images will be had (very close together, because emerald is but feebly doubly refracting). If glass, the images on the card will be single.

Similarly, ruby can at once be distinguished from even the finest garnet or ruby spinel, as the last two are singly refracting. So, too, are glass imitations of ruby and ruby doublets (which consist of glass and garnet). This test cannot injure the stone, it may be applied to mounted stones, and it is reliable. For stones of very deep color this test may fail for lack of sufficiently brilliant reflections. In such a case hold the card *beyond* the stone and let the sunlight shine *through* the stone onto the card, observing whether the spots of light are single or double.

The table below gives the necessary information as to which stones show double and which single refraction.

TABLE GIVING CHARACTER OF REFRACTION IN THE PRINCIPAL GEMS

Refraction Single:

- Diamond
- Garnet (all types)
- Spinel
- Opal
- Glass

Refraction Double	*Difference between highest and lowest refractive indices*
Sphene	.084
Zircon	.053
Benitoite	.047
Peridot or chrysolite	.038
Epidote	.031
Tourmaline	.020
Kunzite	.015
Ruby and sapphire	.009
Topaz (precious)	.009
Amethyst and quartz topaz	.009
Emerald and aquamarine	.007
Chrysoberyl	.007

The student should now put into practice the methods suggested in this lesson. Look first for the visible doubling of the lines of the back facets in peridot (or chrysolite); then in zircon; then in some of the less strongly doubly refracting stones; then try the sunlight-card method with genuine stones and with doublets and imitations until you can tell every time whether you are dealing with singly or doubly refracting material. When a stone of unknown indentity comes along, try the method on it and thus assign it as a first step to one or the other class. Other tests will then be necessary to definitely place it.

Differences in Refraction Due to Crystal Form. The difference in behavior toward light of the singly and doubly refracting minerals depends upon the crystal structure of the mineral. All gems whose crystals belong in the cubic system are singly refracting in all directions: In the case of some other systems of crystals the material may be singly refracting in one or in two directions, but doubly refracting in other directions. No attention need be paid to these complications, however, when using the sunlight-card method with a cut stone, for in such a case the light in its course within the stone will have crossed the material in two or more directions, and the separation and consequent doubling of image will be sure to result. For those who wish to study double refraction more in detail, Chapter VI., pages 40–52, of G. F. Herbert-Smith's *Gem-Stones* will serve admirably as a text. As an alternative any text-book on physics will answer.

LESSON IV

ABSORPTION AND DICHROISM

Cause of Color in Minerals. In Lesson III. we saw that many gem materials cause light that enters them to divide and take two paths within the material. Now all transparent materials *absorb* light more or less; that is, they stop part of it, perhaps converting it into heat, and less light emerges than entered the stone. If light of all the rainbow colors (red, orange, yellow, green, blue, violet) is equally absorbed, so that there is the same relative amount of each in the light that comes out as in the light that went into a stone, we say that the stone is a *white* stone; that is, it is not a *colored* stone. If, however, only blue light succeeds in getting through, the rest of the white light that entered being absorbed within, we say that we have a blue stone.

Similarly, the *color* of any transparent material depends upon its relative degree of absorption of each of the colors in white light. That color which emerges most successfully gives its name to the color of the stone. Thus a ruby is red because red light succeeds in passing through the material much better than light of any other color.

Unequal Absorption Causes Dichroism. All that has been said so far applies equally well to both singly and doubly refracting materials, but in the latter sort it is frequently the case, in those directions in which light always divides, that the absorption is not equal in the two beams of light (one is called the ordinary ray and the other the extraordinary ray).

For example, in the case of a crystal of ruby, if white light starts to *cross* the crystal, it not only divides into an ordinary ray and an extra-ordinary ray, but the absorption is different in the two cases, and the two rays emerge of different shades of red. With most rubies one ray emerges purplish-red, the other yellowish-red.

It will at once be seen that if the human eye could distinguish between the two rays, we would have here a splendid method of determining many precious stones. Unfortunately, the eye does not analyze light, but rather blends the effect so that the unaided eye gives but a poor means of telling whether or not a stone exhibits twin colors, or *dichroism*, as it is called. (The term signifies two colors.) A well-trained eye can, however, by viewing a stone in several different positions, note the difference in shade of color caused by the differential absorption.

The Dichroscope. Now, thanks to the scientific workers, there has been devised a relatively simple and comparatively inexpensive instrument called the *dichroscope*, which enables one to tell almost at a glance whether a stone is or is not dichroic. The construction is indicated in the accompanying drawing and description.

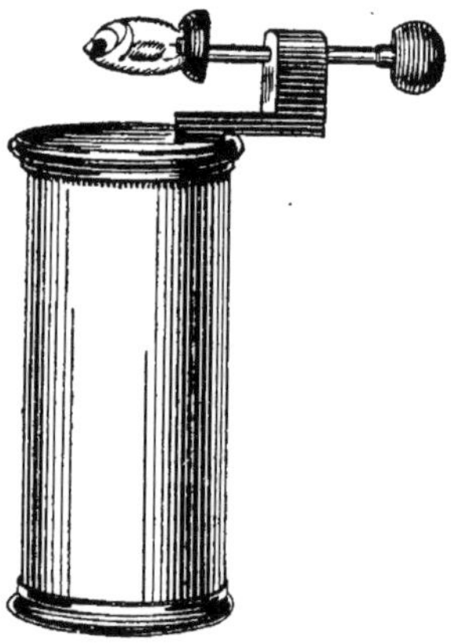

THE DICHROSCOPE

If the observer looks through the lens (A)

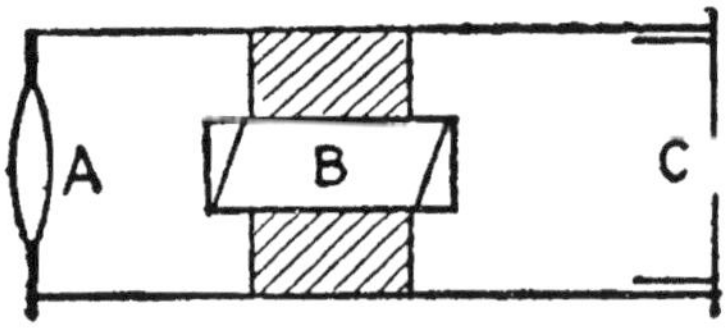

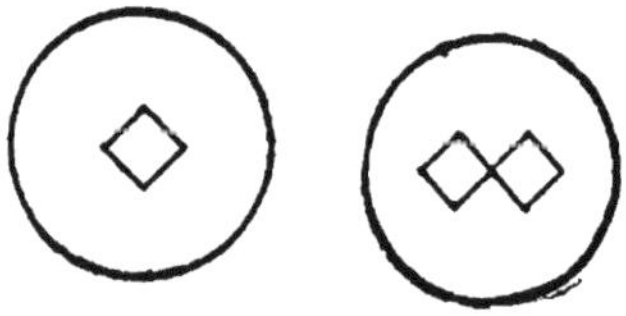

FIG. 3.
A, simple lens; B, piece of Iceland spar with glass prisms on ends to square them up; C, square hole.

FIG. 4.

toward a bright light, as, for example, the sky, he apparently sees two square holes, Fig. 4.

What has happened is that the light passing through the square hole (C of Fig. 3) has divided in passing through the strongly doubly refracting Iceland spar (B of Fig. 3) and two images of the square hole are thus produced.

If now a stone that exhibits dichroism is held in front of the square hole and viewed toward the light, two images of the stone are seen, one due to its ordinary ray (which, as was said above, will have one color), and the other due to its extraordinary ray (which will have a different color or shade of color), thus the color of the two squares will be different.

With a singly refracting mineral, or with glass, or with a doubly refracting mineral when viewed in certain directions of the crystal (which do not yield double refraction) the colors will be alike in the two squares. Thus to determine whether a red stone is or is not a ruby (it might be a garnet or glass or a doublet, all of which are singly refracting and hence can show no dichroism), hold the stone before the hole in the dichroscope and note whether or not it produces twin colors. If there seems to be no difference of shade turn the stone about, as it may have accidentally been placed so that it was viewed along its direction of single refraction. If there is still no dichroism it is not a ruby. (*Note.*—Scientific rubies exhibit dichroism as well as natural ones, so this test will not distinguish them.)

A dichroscope may be had for from seven to ten dollars, according to the make, and everyone who deals in colored stones should own and use one.

Not all stones that are doubly refracting exhibit dichroism. White stones of course cannot exhibit it even though doubly refracting, and some colored stones, though strongly doubly refracting, do not exhibit any noticeable dichroism. The zircon, for example, is strongly doubly refracting, but shows hardly any dischroism.

The test is most useful for emerald, ruby,

sapphire, tourmaline, kunzite and alexandrite, all of which show marked dichroism.

It is of little use to give here the twin colors in each case as the shades differ with different specimens, according to their depth and type of color. The deeper tinted stones of any species show the effect more markedly than the lighter ones.

The method is rapid and easy—it can be applied to mounted stones as well as to loose ones, and it cannot injure a stone. The student should, if possible, obtain the use of a dichroscope and practice with it on all sorts of stones. He should especially become expert in distinguishing between rubies, sapphires, and emeralds, and their imitations. The only imitation (scientific rubies and sapphires are not here classed as imitations), which is at all likely to deceive one who knows how to use the dichroscope is the emerald triplet, made with real (but pale) beryl above and below, with a thin strip of green glass between. As beryl is doubly refracting to a small degree, and dichroic, one might perhaps be deceived by such an imitation if not careful. However, the amount of dichroism would be less in such a case than in a true emerald of as deep a color.

Those who wish to study further the subject of dichroism should see *Gem-Stones*, by G. F. Herbert-Smith, Chapter VII., pp. 53–59, or see *A Handbook of Precious Stones*, by M. D. Rothschild, Putnam's, pp. 14–16.

LESSON V

SPECIFIC GRAVITY

THE properties so far considered as serving to distinguish precious stones have all depended upon the behavior of the material toward light.

These properties were considered first because they afford, to those acquainted with their use, very rapid and sure means of classifying precious stones.

Density of Minerals. We will next consider an equally certain test, which, however, requires rather more time, apparatus, and skill to apply.

Each kind of precious stone has its own *density*. That is, if pieces of *different* stones were taken all of the same *size*, the *weights* would differ, but like-sized pieces of one and the same material always have the same weight. It is the custom among scientists to compare the densities of substances with the density of water. The number which expresses the relation between the density of any substance and the density of water is called the *specific gravity* number of the substance. For example, if, size for size, a material, say diamond, is 3.51 times as heavy as water, its *specific gravity* is 3.51. It will be seen that since each substance always has, when pure, the same specific gravity, we have here a means of distinguishing precious stones. It is very seldom, if ever, the case that we find any two precious stones of the same specific gravity. A few stones have nearly the same specific gravities, and in such cases it is well to apply other tests also. *In fact one should always make sure of a stone by seeing that two or three different tests point to the same species.*

We must next find out how to determine the specific gravity of a precious stone. If the shape of a stone were such that the volume could be readily calculated, then one could easily compare the weight with the volume or with the weight of the same volume of water, and thus get the specific gravity (for a specific gravity number really tells how much heavier a piece of material is than the same volume of water).

Unfortunately the form of most precious stones is such that it would be very difficult to calculate the volume from the measurements, and the latter would be hard to make accurately with small stones. To avoid these difficulties the following ingenious method has been devised:

If a stone is dropped into water it pushes aside, or *displaces*, a body of water exactly equal in volume to itself. If the water thus displaced were caught and weighed, and the weight of the stone then divided by the weight of the water displaced, we would have the specific gravity number of the stone.

This is precisely what is done in getting the

specific gravity of small stones. To make sure of getting an accurate result for the weight of water displaced the following apparatus is used.

The Specific Gravity Bottle. A small flask-like bottle (see Fig. 5) is obtained. This has

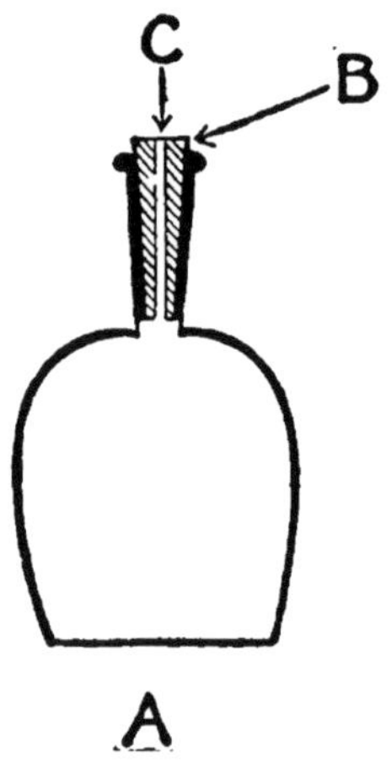

FIG. 5
A, Flask-like Bottle; B, Indicates Ground Glass Stopper; C, Shows Hole Drilled through Stopper

a tightly fitting *ground* glass stopper (B). The stopper has a small hole (C) drilled through it lengthwise. If the bottle is filled with water, and the stopper dropped in and tightened, water will squirt out through the small hole in the stopper. On wiping off stopper and bottle we have the bottle *exactly full* of water. If now the stopper is removed, the stone to be tested (which must of course be smaller than the neck of the bottle) dropped in, and the stopper replaced, exactly as much water will squirt out as is equal in volume to the stone that was dropped in.

If we had weighed the full bottle with the stone *on the pan beside it*, and then weighed the bottle with the stone *inside it* we could now, by subtracting the last weight from the first, find out how much the water, that was displaced, weighed. This is precisely the thing to do. The weight of the stone being known we now have merely to divide the weight of the stone by the weight of the displaced water, and we have the specific gravity number. Reference to a table of specific gravities of precious stones will enable us to name our stone. Such a table follows this lesson.

A Sample Calculation. The actual performance of the operation, if one is skilled in weighing, takes less time than it would to read this description. At first one will be slow, and perhaps one should read and re-read this lesson, making sure that all the ideas are clear before trying to put them in practice.

A sample calculation may help make the matter clearer, so one is appended:

Weight of bottle + stone (outside)..........	= 53.51 carats
Weight of bottle + stone (inside)...........	= 52.51 carats
Weight of water displaced	= 1.00 carat
Weight of stone	= 3.51 carats

$$\text{Specific gravity} = \frac{\text{Weight of stone}}{\text{Weight of water}} = \frac{3.51}{1.00} = 3.51 \text{ Sp. g.}$$

In this case the specific gravity being 3.51, the stone is probably diamond (see table), but might be precious topaz, which has nearly the same specific gravity.

It is assumed that the jeweler will weigh in carats, and that his balance is sensitive to .01 carat. With such a balance, and a specific gravity bottle (which any scientific supply house will furnish for less than $1) results sufficiently accurate for the determination of precious stones may be had if one is careful to exclude air bubbles from the bottle, and to wipe the outside of the bottle perfectly dry before each weighing. The bottle should never be held in the warm hands, or it will act like a thermometer and expand the water up the narrow tube in the stopper, thus leading to error. A handkerchief may be used to grasp the bottle.

Table of Specific Gravities of the Principal Gem Materials

Beryl (Emerald)	2.74
Chrysoberyl (Alexandrite)	3.73
Corundum (Ruby, sapphire, "Oriental topaz")	4.03
Diamond	3.52
Garnet (Pyrope)	3.78
" (Hessonite)	3.61
" (Demantoid, known in the trade as "Olivine")	3.84
" (Almandite)	4.05
Opal	2.15
Peridot	3.40
Quartz (Amethyst, common topaz)	2.66
Spinel (Rubicelle, Balas ruby)	3.60
Spodumene (Kunzite)	3.18
Topaz (precious)	3.53
Tourmaline	3.10
Turquoise	2.82
Zircon, lighter variety	4.20
" heavier variety	4.69

For a more complete and scientific discussion of specific gravity determination see *Gem-Stones*, by G. F. Herbert-Smith, Chapter VIII., pp. 63–77; or see, *A Handbook of Precious Stones*, by M. D. Rothschild, pp. 21–27, for an excellent account with illustrations; or see any physics text-book.

LESSON VI

SPECIFIC GRAVITY DETERMINATIONS

Weighing a Gem in Water. In the previous lesson it was seen that the identity of a precious stone may be found by determining its specific gravity, which is a number that tells how much heavier the material is than a like volume of water. It was not explained, however, how one would proceed to get the specific gravity of a stone too large to go in the neck of a specific gravity bottle. In the latter case we resort to another method of finding how much a like volume of water weighs. If the stone, instead of being dropped into a perfectly full bottle of water (which then overflows), be dropped into a partly filled glass or small beaker of water, just as much water will be displaced as though the vessel were full, and it will be displaced *upward* as before, for lack of any other place to go. Consequently its weight will tend to buoy up or float the stone by trying to get back under it, and the stone when in water will weigh less than when in air. Anyone who has ever pulled up a small anchor when out fishing from a boat will recognize at once that this is the case, and that as the anchor emerges from the water it seems to suddenly grow heavier. Not only does the stone weigh less when in the water, but it weighs exactly as much less as the weight of the water that was displaced by the stone (which has a volume equal to the volume of the stone). If we weigh a stone first in the air, as usual, and then in water (where it weighs less), and then subtract the weight in water from the weight in air we will have the *loss of weight in water*, and this equals the *weight of an equal volume of water*, which is precisely what we got by our bottle method.

We now need only divide the weight in air

by the loss of weight in water, and we shall have the specific gravity of the stone.

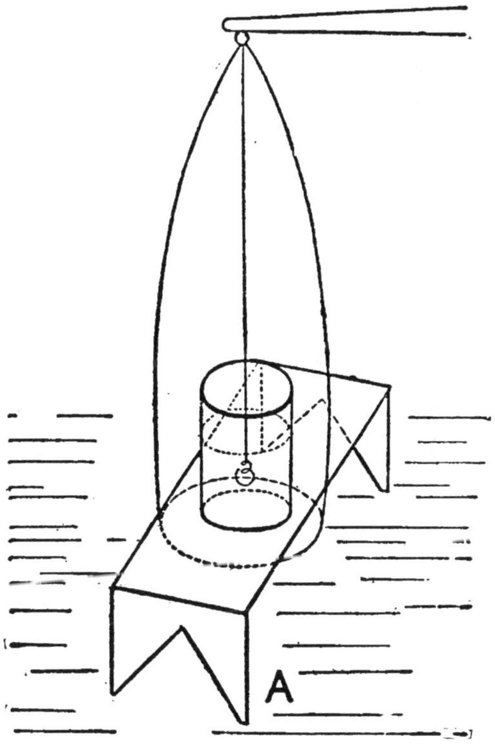

FIG, 6.

To actually weigh the stone in water we must use a fine wire to support the stone. We must first find how much this wire itself weighs (when attached by a small loop to the hook that supports the balance pan and trailing partly in the water, as will be the case when weighing the stone in water). This weight of the wire must of course be deducted to get the true weight of the stone in water. The beaker of water is best supported by a small table that stands over the balance pan. One can easily be made out of the pieces of a cigar box. (See Fig. 6.)

The wire that is to support the stone should have a spiral at the bottom in which to lay the gem, and this should be so placed that the latter will be completely submerged at all times, but not touching bottom or sides of the beaker.

Example of data, and calculation, when getting specific gravity by the method of weighing in water:

Weight of stone..........................	= 4.02 carats
Weight of stone (plus wire) in water	= 3.32 carats
Weight of wire..........................	= .30 carat
True weight of stone in water	= 3.02 carats
Loss of weight in water....................	= 1.00 carat

$$\text{Specific gravity} = \frac{\text{Weight of stone}}{\text{Loss in water}} = \frac{4.02}{1.00} = 4.02$$

Here the specific gravity, 4.02 would indicate some corundum gem (ruby or sapphire), and the other characters would indicate at once which it was.

The student who means to master the use of the two methods given in Lessons V. and VI. should proceed to practice them with stones of known specific gravities until he can at least get the correct result to the first decimal place. It is not to be expected that accurate results can be had in the second decimal place, with the balances usually available to jewelers. When the learner can determine specific gravities with some certainty he should then try unknown gems.

The specific gravity method is of especial value in distinguishing between the various colorless stones, as, for example, quartz crystal, true white topaz, white sapphire, white or colorless beryl, etc. These are all doubly refractive, have no color, and hence no dichroism, and unless one has a refractometer to get the refractive index, they are difficult to distinguish. The specific gravities are very different, however, and readily serve to distinguish them. It should be added that the synthetic stones show the same specific gravities as their natural counterparts, so that this test does not serve to detect them.

Where many gems are to be handled and separated by specific gravity determinations, perhaps the best way to do so is to have several liquids of known specific gravity and to see what stones will float and what ones will sink in the liquids. Methylene iodide is a heavy liquid (sp. g. 3.32), on which a "quartz-topaz," for example, sp. g. 2.66, would float, but a true topaz, sp. g. 3.53, would sink in it. By diluting methylene iodide with benzol (sp. g. 0.88) any specific gravity that is desired may be had (between the two limits 0.88 and 3.32). Specimens of known specific gravity are used with such liquids and their behavior (as to whether they sink or float, or remain suspended in the liquid,) indicates the specific gravity of the liquid. An unknown stone may then be used and its behavior noted and compared with that of a known specimen, whereby one can easily find out whether the unknown is heavier or lighter than the known sample.

An excellent account of the detail of this method is given in G. F. Herbert-Smith's *Gem-Stones*, pages 64–71, of Chapter VIII., and various liquids are there recommended. It is doubtful if the practical gem dealer would find these methods necessary in most cases. Where large numbers of many different unknown gems have to be determined it would pay to prepare, and standardize, and use such solutions.

LESSON VII

LUSTER AND OTHER REFLECTION EFFECTS

BY the term *luster* we refer to the manner and degree in which light is reflected from the *surface* of a material. Surfaces of the same material, but of varying degrees of smoothness would, of course, vary in the vividness of their luster, but the type of variation that may be made use of to help distinguish gems, depends upon the character of the material more than upon the degree of smoothness of its surface. Just as silk has so typical a luster that we speak of it as silky luster, and just as pearl has a pearly luster, so certain gems have peculiar and characteristic luster. The diamond gives us a good example. Most diamond dealers distinguish between real and imitation diamonds at a glance by the character of the luster. That is the chief, and perhaps the only property, that they rely upon for deciding the genuineness of a diamond, and they are fairly safe in so doing, for, with the exception of certain artificially decolorized zircons, no gem stone is likely to deceive one who is familiar with the luster of the diamond. It is not to be denied that a fine white zircon, when finely cut, may deceive even one who is familiar with diamonds. The author has fooled many diamond experts with an especially fine zircon, for the luster of zircon does approach, though it hardly equals, that of the diamond. Rough zircons are frequently mistaken for diamonds by diamond prospectors, and even by pickers in the mines, so that some care should be exercised in any suspicious case, and one should not then rely solely on the luster. However, in most cases in the trade there is almost no chance of the unexpected presence of a zircon and the luster test is usually sufficient to distinguish the diamond. (Zircons are strongly doubly refractive, as was said in Lesson III. on Double Refraction, and with a lens the doubling of the back lines may be seen.)

Adamantine Luster. The luster of a diamond is called *adamantine* (the adjective uses the Greek name for the stone itself). It is keen and cold and glittering, having a metallic suggestion. A very large per cent. of the light that falls upon the surface of a diamond at any low angle is reflected, hence the keenness of its luster. If a diamond and some other white stone, say a white sapphire, are held so as to reflect at the same time images of an incandescent light into the eye of the observer, such a direct comparison will serve to show that much more light comes to the eye from the diamond surface than from the sapphire surface. The image of the light filament, as seen from the diamond, is much keener than as seen from the sapphire. The same disparity would exist between the diamond and almost any other stone. Zircon comes nearest to having adamantine luster of any of the other gems. The green garnet that is called "olivine" in the trade also approaches diamond in luster, hence the name "demantoid," or diamond like, sometimes applied to it.

Vitreous Luster. The other stones nearly all have what is called *vitreous* luster (literally, glass like), yet owing to difference of hardness, and consequent minute differences in fineness of surface finish, the keenness of this vitreous luster varies slightly in different stones, and a trained eye can obtain clues to the identity of certain stones by means of a consideration of the luster. Garnets, for example, being harder than glass, take a keener polish, and a glance at a doublet (of which the hard top is usually garnet and the base of glass) will show that the light is better reflected from the garnet part of the top slope than from the glass part. This use of luster affords the quickest and surest means of detecting a doublet. One can even tell a doublet inside a show window, although the observer be outside on the sidewalk, by moving to a position such that a reflection from the top slope of the stone is to be had. When a doublet has a complete garnet top no such direct comparison can be had, but by viewing first the top luster, and then the back luster, in rapid succession, one can tell whether or not the stone is a doublet.

Oily Luster. Certain stones, notably the peridot (or chrysolite) and the hessonite (or cinnamon stone), have an oily luster. This is possibly due to reflection of light that has penetrated the surface slightly and then been reflected from disturbed layers beneath the surface. At any rate, the difference in luster may be made use of by those who have trained their eyes to appreciate it. Much practice will be needed before one can expect to tell at a glance when he has a peridot (or chrysolite) by the luster alone, but it will pay to spend some spare time in studying the luster of the various stones.

A true, or "precious" topaz, for example,

may be compared with a yellow quartz-topaz, and owing to the greater hardness of the true topaz, it will be noted that it has a slightly keener luster than the other stone, although both have vitreous luster. Similarly the corundum gems (ruby and sapphire), being even harder than true topaz, take a splendid surface finish and have a very keen vitreous luster.

Turquoise has a dull waxy luster, due to its slight hardness. Malachite, although soft, has, perhaps because of its opacity, a keen and sometimes almost metallic luster.

One may note the luster rapidly, without apparatus and without damage to the stone. We thus have a test which, while it is not conclusive except in a very few cases, will supplement and serve to confirm other tests, or perhaps, if used at first, will suggest what other tests to apply.

Another optical effect that serves to distinguish some stones depends upon the reflection of light from within the material due to a certain lack of homogeneity in the substance.

Cause of Color in the Opal. Thus the opal is distinguished by the prismatic colors that emerge from it owing to the effect of thin layers of material of slightly different density, and hence of different refractive index from the rest of the material. These thin films act much as do soap-bubble films, to interfere with light of certain wave lengths, but to reflect certain other wave lengths and hence certain colors.

Again, in some sapphires and rubies are found minute, probably hollow, tube-like cavities, arranged in three sets in the same positions as the transverse axes of the hexagonal crystal. The surfaces of these tubes reflect light so as to produce a six-pointed star effect, especially when the stone is properly cut to a high, round cabochon form, whose base is parallel to the successive layers of tubes.

Starstones, Moonstones, Cat's-eye. In the moonstone we have another sort of effect, this time due to the presence of hosts of small twin crystal layers that reflect light so as to produce a sort of moonlight-on-the-water appearance *within* the stone when the latter is properly cut, with the layers of twin crystals parallel to its base. Ceylon-cut moonstones are frequently cut to save weight, and may have to be recut to properly place the layers so that the effect may be seen equally over all parts of the stone, as set.

Cat's-eye and tiger's-eye owe their peculiar appearance to the presence, within them, of many fine, parallel, silky fibers. The quartz cat's-eye was probably once an asbestos-like mineral, whose soft fibers were replaced by quartz in solution, and the latter, while giving its hardness to the new mineral, also took up the fibrous arrangement of the original material. The true chrysoberyl cat's-eye also has a somewhat similar fibrous or perhaps tubular strue-ture. Such stones, when cut *en cabochon*, show a thin sharp line of light running across the center of the stone (when properly cut with the base parallel to the fibers). This is due to reflection of light from the surfaces of the parallel fibers. The line of light runs perpendicularly to the fibers.

In these cases (opals, starstones, moonstones, and cat's-eyes) the individual stone is usually easily distinguished from other kinds of stones by its peculiar behavior towards light. However, it must be remembered that other species than corundum furnish starstones (amethyst and other varieties of quartz, for example), so that it does not follow that any starstone is a corundum gem. Also the more valuable chrysoberyl cat's-eye may be confused with the cheaper quartz cat's-eye unless one is well acquainted with the respective appearances of the two varieties. Whenever there is any doubt other tests should be applied.

For further account of luster and other types of reflection effects see *Gem-Stones*, by G. F. Herbert-Smith, Chapter V., pp. 37–39, or *A Handbook of Precious Stones*, M. D. Rothschild, pp. 17, 18.

LESSON VIII

HARDNESS

ANOTHER property by means of which one may distinguish the various gems from each other is *hardness*. By hardness is meant the ability to resist scratching. The term "hardness" should not be taken to include toughness, yet it is frequently so understood by the public. Most hard stones are more or less brittle and would shatter if struck a sharp blow. Other hard stones have a pronounced *cleavage* and split easily in certain directions. True hardness, then, implies merely the ability to resist abrasion (*i. e.*, scratching).

Now, not only is hardness very necessary in a precious stone in order that it may *receive* and *keep* a fine polish, but the degree in which it possesses hardness as compared with other materials of known hardness may be made use of in identifying it.

No scale of *absolute* hardness has ever come into general use, but the mineralogist Mohs many years ago proposed the following *relative* scale, which has been used very largely:

Mohs's Scale of Hardness. Diamond, the hardest of all gems, was rated as 10 by Mohs. This rating was purely arbitrary. Mohs might have called it 100 or 1 with equal reason. It was merely in order to represent the different degrees of hardness by numbers, that he picked out the number 10 to assign to diamonds. Sapphire (and ruby) Mohs called 9, as being next to diamond in hardness. True topaz (precious topaz) he called 8. Quartz (amethyst and quartz "topaz") was given the number 7. Felspar (moonstone) was rated 6, the mineral apatite 5, fluorspar 4, calcite 3, gypsum 2, and talc 1.

It may be said here that any mineral in this series, that is of higher number than any other, will scratch the other. Thus diamond (10) will scratch all the others, sapphire (9) will scratch any but diamond, topaz (8) will scratch any but diamond and sapphire, and so on.

It must not be thought that there is any regularity in the degrees of hardness as expressed by these numbers. The intervals in hardness are by no means equal to the differences in number. Thus the interval between diamond and sapphire, although given but one number of difference, is probably greater than that between sapphire (9) and talc (1). The numbers thus merely give us an order of hardness. Many gem minerals are, of course, missing from this list, and most of the minerals from 5 down to 1 are not gem minerals at all. Few gem materials are of less hardness than 7, for any mineral less hard than quartz (7) will inevitably be worn and dulled in time by the ordinary road dust, which contains much powdered quartz.

In testing a gem for hardness the problem consists in finding out which of the above minerals is most nearly equal in hardness to the unknown stone. Any gem that was approximately equal in hardness to a true topaz (8) would also be said to be of hardness 8. Thus spinel is of about the same hardness as topaz and hence is usually rated as 8 in hardness. Similarly opal, moonstone, and turquoise are of about the same hardness as felspar and are all rated 6.

Frequently stones will be found that in hardness are between some two of Mohs's minerals. In that case we add one half to the number of the softer mineral; thus, peridot, benitoite, and jade (nephrite) are all softer than quartz (7) but harder than felspar (6); hence we say they are 6½ in hardness. Beryl (aquamarine and emerald), garnet (almandine), and zircon are rated 7½ in hardness, being softer than true topaz but harder than quartz. A table of the hardness of most of the commonly known gem-stones follows this lesson.

Having now an idea of what hardness means and how it is expressed, we must next inquire how one may make use of it in identifying unknown gems.

How to Apply the Hardness Test. In the first place, it is necessary to caution the beginner against damaging a fine gem by attempting to test its hardness in any but the most careful manner. The time-honored file test is really a hardness test and serves nicely to distinguish genuine gems, of hardness 7 or above, from glass imitations. A well-hardened steel file is of not quite hardness 7, and glass of various types while varying somewhat averages between 5 and 6. Hence, glass imitations are easily attacked by a file. To make the file test use only a *very fine* file and apply it with a light but firm pressure lengthwise along the girdle (edge) of the unset stone. If damage results it will then be almost unnoticeable.

Learn to know the *feel* of the file as it takes hold of a substance softer than itself. Also learn the *sound*. If applied to a hard stone a file will slip on it, as a skate slips on ice. It will not take hold as upon a softer substance.

If the stone is set, press a sharp corner of a broken-ended file gently against a *back* facet, preferably high up toward the girdle, where any damage will not be visible from the front, and move the file very slightly along the surface, noting by the *feel* whether or not it takes hold and also looking with a lens to see if a scratch has been made. Do not mistake a line of steel, left on a slightly rough surface, for a true scratch. Frequently on an unpolished girdle of real gem material the file will leave a streak of steel. Similarly when using test minerals in accordance with what follows do not mistake a streak of powder from the yielding test material, for a true scratch in the material being tested. The safe way is to wipe the spot thus removing any powder. A true scratch will, of course, persist.

A doublet, being usually constructed of a garnet top and a glass back, may resist a file at the girdle if the garnet top covers the stone to the girdle, as is sometimes the case, especially in the smaller sizes. In this case the back must be tested.

One should never pass a file rudely across the corners or edges of the facets on any stone that may be genuine, as such treatment really amounts to a series of light hammer blows, and the brittleness of most gem stones would cause them to yield, irrespective of their hardness. It should be remembered that some genuine stones are softer than a file, so that it will not do to reject as worthless any material that is attacked by a file. Lapis lazuli (5), sphene (5), opal (6), moonstone (6), amazonite (6), turquoise (6), peridot (6½), demantoid garnet (6½) (the "olivine" of the trade), and jade (nephrite) (6½), are all more or less attacked by a file.

TABLE OF HARDNESS OF THE PRINCIPAL GEM-STONES

10.	Diamond.
9½.	(Carborundum.)
9.	Sapphire and ruby (also all the color varieties of sapphire).
8½.	Chrysoberyl (alexandrite).
8.	True topaz and spinel (rubicelle, balas ruby).
7½.	Emerald, aquamarine, beryl, Morganite, zircon (jacinth and true hyacinth and jargoon), almandine garnet.
7¼.	Pyrope garnet (Arizona ruby, cape ruby), hessonite garnet (cinnamon stone), tourmaline (various colors vary from 7 to 7½), kunzite (7+)
7.	Amethyst, various quartz gems, quartz "topaz," jade (jadeite).
6½.	Peridot (chrysolite), demantoid garnet ("olivine"), jade (nephrite).
6.	Opal, moonstone, turquoise.
5.	Lapis lazuli.

LESSON IX

HARDNESS—*Continued*

Minerals Used in Testing Hardness. For testing stones that are harder than a file the student should provide himself with the following set of materials:

1. A small crystal of carborundum. (Most hardware stores have specimen crystals as attractive advertisements of carborundum as an abrasive material, or the Carborundum Co., Niagara Falls, N. Y., will supply one.)

2. A small crystal of sapphire (not of gem quality, but it should be transparent and compact. A palc or colorless Montana sapphire can be had for a few cents of any mineral dealer).

3. A small *true topaz* crystal. (The pure white topaz of Thomas Mountain, Utah, is

excellent; or white topaz from Brazil or Japan or Mexico or Colorado will do. Any mineral house can furnish small crystals for a few cents when not of specially fine crystallization.)

4. A small quartz crystal. (This may be either amethyst or quartz-topaz or the common colorless variety. The fine, sharp, colorless crystals from Herkimer County, N. Y., are excellent. These are very inexpensive.)

5. A fragment of a crystal of felspar. (Common orthoclase felspar, which is frequently of a brownish pink or flesh color, will do.)

These five test stones represent the following degrees of hardness:

1. Carborundum is harder than any gem material but diamond. It will scratch sapphire and ruby, which are rated 9 in hardness, hence we may call carborundum 9½ if we wish. It is, however, very much softer than diamond, and the latter will scratch it upon the slightest pressure.

2. Sapphire, of hardness 9, scratching any gem material except diamond.

3. True topaz, of hardness 8. It is scratched by sapphire (and, of course, ruby), also by chrysoberyl (which is hence rated 8½), but scratches most other stones. Spinel (which is also rated as 8 in hardness) is really a bit harder than topaz.

4. Quartz, of hardness 7, and scratched by all the previous stones but scratching those that were listed above as of less hardness than a file.

5. Felspar, of hardness 6, hence slightly softer than a file and yielding to it, but scratching the stones likewise rated as 6 when applied forcibly to them. Also scratching stones rated as less than 6 on slight pressure.

We must next consider how these minerals may be safely used upon gem material. Obviously it would be far safer to use them upon rough gem material than upon cut stones. However, with care and some little skill, one may make hardness tests without particular danger to fine cut material.

The way to proceed is to apply the cut stone (preferably its girdle, or if that is so set as not to be available, a corner where several facets meet) gently to the flat surface of one of the softer test stones, drawing it lightly along the surface and noting the *feel* and looking to see if a scratch results. If the test stone is scratched try the next harder test stone similarly. *Do not attempt to use the test stone upon any valuable cut stone.* Proceed as above until the gem meets a test stone that it does not attack. Its hardness is then probably equal to the latter and perhaps if pressed forcibly against it a slight scratch would result, but it is not advisable to resort to heavy pressure. A light touch should be cultivated in this work. Having now an indication as to the hardness of the unknown gem look up in the table of the previous lesson those gems of similar hardness and then by the use of some of the tests already given decide which of the stones of that degree of hardness you have. *Never rely upon a single test in identifying a gem.*

For further study of hardness and its use in testing gems see *Gem-Stones*, G. F. Herbert-Smith, Chap IX., pp. 78–81, and table on p. 305; or see *A Handbook of Precious Stones*, Rothschild, pp. 19, 20, 21.

LESSON X

DISPERSION

ANOTHER property which may be made use of in deciding the identity of certain gems is that called *dispersion.* We have seen in Lesson II. that light in entering a stone from the air changes its path (refraction), and in Lesson III. it was explained that many minerals cause light that enters them to divide and proceed along two different paths (double refraction). Now it is further true that light of the various colors (red, orange, yellow, green, blue, and violet) is refracted variously—the violet being bent most sharply, the red least, and the other colors to intermediate degrees. The cut (Fig. 7) represents roughly and in an exaggerated manner the effect we are discussing.

Now in a cut stone this separation of light

of different colors, or dispersion of light, as it is called, results in the reflection of each of the colors separately from the steep sloping back facets of the stone. If almost any clear, colorless faceted stone is placed in the sunlight

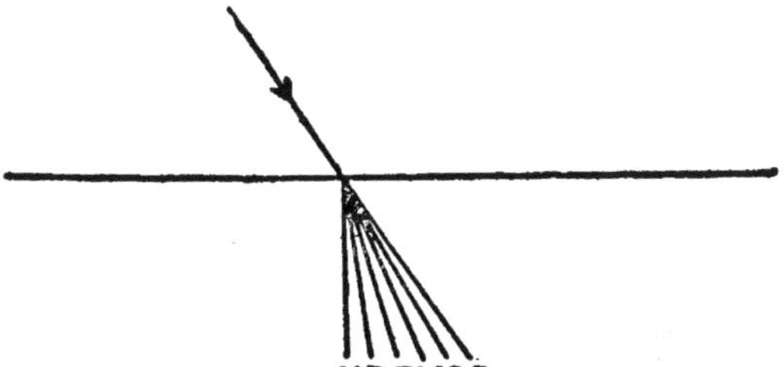

FIG. 7.

and a card held before it to receive the reflections, it will be seen that rainbow-like reflections appear on the card. These *spectra*, as they are called, are caused by the dispersion of light. With a diamond the spectra will be very brilliant and of vivid coloring, and the red will be widely separated from the blue. With white sapphire or white topaz, or with rock crystal (quartz), the spectra will be less vivid—they will appear in pairs (due to the double refraction of these minerals), and the red and blue will be near together (*i. e.*, the spectra will be short). This shortness in the latter cases is due to the small dispersive power of the three minerals mentioned. Paste (lead glass) gives fairly vivid spectra, and they are single like those from diamond, as glass is singly refracting. The dispersion of the heavy lead glass approaches that of diamond. The decolorized zircon (jargoon) has a dispersion well up toward that of diamond and gives fairly vivid spectra on a card, but they are double, as zircon is doubly refracting. Sphene (a gem rarely seen in the trade) and the demantoid garnet (a green gem often called "olivine" in the trade) both have very high dispersive power, exceeding the diamond in this respect. As they are both colored stones (sphene is usually yellowish, sometimes greenish or brown), the vividness of their color-play is much diminished by absorption of light within them. So also the color-play of a deeply colored fancy diamond is diminished by absorption.

Dispersion as a Test of the Identity of a Gem. We may now consider how an acquaintance with the dispersive powers of the various stones can be used in distinguishing them. If a stone has high dispersive power it will exhibit "fire," as it is called—*i. e.*, the various colors will be so widely separated within the stone, and hence reflected out so widely separated, that they will fall on the eye (as on the card above) in separate layers, and vivid flashes of red or yellow or other colors will be seen. Such stones as the white sapphire (and others of small dispersion), however, while separating the various colors appreciably as seen reflected on a card, do not sufficiently separate them to produce the "fire" effect when the light falls on the eye. This is because the various colors, being very near together in this case, cross the eye so rapidly, when the stone is moved, that they blend their effect and the eye regards the light that thus falls upon it as white. We have here a ready means of distinguishing the diamond from most other colorless gems. The trained diamond expert relies (probably unconsciously) upon the dispersive effect (or "fire") nearly as much as upon the adamantine luster, in telling at a glance whether a stone is or is not a diamond. Of all colorless stones, the only one likely to mislead the expert in this respect is the whitened zircon (jargoon), which has almost adamantine luster and in addition nearly as high dispersive power as diamond. However, zircon is doubly refracting (strongly so), and the division of the spectra which results (each facet producing two instead of only one) weakens the "fire" so that even the best zircon is a bit "sleepy" as compared with even an ordinary diamond.

In addition to providing a ready means of identifying the diamond, a high degree of dispersion in a stone of pronounced color would lead one to consider sphene, demantoid garnet (if green), and zircon (which might be reddish, yellowish, brown, or of other colors), and if the stone did not agree with these in its other properties one should suspect *glass*.

A good way to note the degree of dispersion, aside from the sunlight-card method, is to look at the stone from the back while holding it up to the light (daylight). Stones of high dispersive power will display vivid color play in this position. Glass imitations of rubies, emeralds, amethysts, etc., will display altogether too much dispersion for the natural gems.

In Chap. III., p. 20, of G. F. Herbert-Smith's *Gem-Stones*, a brief account of dispersion is given. College text-books on physics also treat of it, and the latter give an account of how dispersion is measured and what is meant by a coefficient of dispersion. Most gem books say little about it, but as we have seen above, a knowledge of the matter can, when supplemented by other tests, be applied practically in distinguishing gems.

LESSON XI

COLOR

IN reserving to the last the property of *color*, which many dealers in gems use first when attempting to identify a precious stone, I have sought to point out the fact that a determination based solely upon color is very likely to be wrong. So many mineral species are found in so many different colors that to attempt to identify any mineral species by color alone is usually to invite disaster. The emerald, alone among gems, has, when of fine color, a hue that is not approached by any other species. The color of the grass in the springtime fitly describes it. Yet even here the art of man has so closely counterfeited in glass the green of the emerald that one cannot be sure of his stone by color alone. As was suggested earlier in these lessons, the writer has several times recently had occasion to condemn as glass imitations stones for which high prices had been paid as genuine emeralds, those who sold them having relied solely upon a trained eye for color.

Confusion of Gems Due to Similarity of Color. The same tendency to rely upon color causes many in the trade to call all yellow stones "topaz" whether the species be corundum (oriental topaz), true topaz (precious topaz), citrine quartz (quart topaz), heliodor (yellow beryl), jacinth (yellow zircon), or what not.

Similarly the public calls all red stones ruby. Thus we have "cape ruby" and "Arizona ruby" (pyrope garnet), "spinel ruby" (more properly ruby spinel), "Siam ruby" (very dark red corundum), "Ceylon ruby" (pale pinkish corundum), rubellite (pink tourmaline), and lastly Burmah ruby (the fine blood-red corundum).

While it is true that color, unless skillfully estimated and wisely used in conjunction with other properties, is a most unreliable guide, yet when thus used, it becomes a great help and serves sometimes to narrow down the chase, at the start, to a very few species. To thus make use of it requires an actual acquaintance with the various gem materials, in their usual colors and shades and an eye trained to note and to remember minute differences of tint and shade. The suggestions which follow as to usual colors of mineral species must then be used only with discretion and after much faithful study of many specimens of each of the species.

Let us begin with the beginning color of the visible spectrum, red, and consider how a close study of shades of red can help in distinguishing the various red stones from each other. In the first place we will inquire what mineral species are likely to furnish us with red stones. Omitting a number of rare minerals, we have (1) corundum ruby, (2) garnet of various types, (3) zircon, (4) spinel, (5) tourmaline. These five minerals are about the only common species which give us an out-and-out red stone. Let us now consider the distinctions between the reds of these different species. The red of the ruby, whether dark (Siam type), blood red (Burmah type), or pale (Ceylon), is more pleasing usually than the red of any of the other species. Viewed from the back of the stone (by transmitted light) it is still pleasing. It may be purplish, but is seldom orange red. Also, owing to the dichroism of the ruby the red is variable according to the changing position of the stone. It therefore has a certain life and variety not seen in any of the others except perhaps in red tourmaline, which, however, does not approach ruby in fineness of red color.

Red Stones of Similar Shades. The garnet, on the other hand, when of fire-red hue, is darker than any but the Siam ruby. It is also more inclined to orange red or brownish red—and the latter is especially true when the stone is seen against the light (by transmitted light). Its color then resembles that of a solution of "iron" such as is given as medicine. The so-called "almandine" garnets (those of purplish red tint) do not equal the true ruby in brightness of color and when held up to the light show more prismatic colors than the true ruby, owing to the greater dispersion of garnet. The color also lacks variety (owing to lack of dichroism). While a fine garnet may make a fair-looking "ruby" when by itself, it looks inferior and dark when beside a fine ruby. By artificial light, too, the garnet is dark as compared with the true ruby, and the latter shows its color at a distance much more strongly than the garnet.

The red zircon, or true hyacinth, is rare. (Many hessonite garnets are sold as hyacinths in the trade. These are usually of a brownish red.) The red of the hyacinth is never equal to that of the ruby. It is usually more somber, and a bit inclined to a brownish cast. The dispersion of zircon, too, is so large (about 87 per cent. of that of diamond) that some little "color-play" is likely to appear along with the intrinsic color. The luster too is almost adamantine while that of ruby is softer and vitreous. Although strongly doubly refracting, the hyacinth shows scarcely any dichroism and thus lacks variety of color. Hence a trained eye will at once note these differences and not confound the stone with ruby.

Spinels, when red, are almost always more yellowish or more purplish than fine corundum rubies. They are also singly refracting and hence exhibit no dichroism and therefore lack variety of color as compared with true ruby. Some especially fine ones, however, are of a good enough red to deceive even jewelers of experience, and one in particular that I have in mind has been the rounds of the stores and has never been pronounced a spinel, although several "experts" have insisted that it was a scientific ruby. The use of a dichroscope would have saved them that error, for the stone is singly refracting. Spinels are usually clearer and more transparent than garnets and show their color better at a distance or when in a poor light.

Tourmaline of the reddish variety (rubellite) is seldom of a deep red. It is more inclined to be pinkish. The dichroism of tourmaline is stronger than that of ruby and more obvious to the unaided eye. The red of the rubellite should not deceive anyone who has ever seen a fine corundum ruby.

YELLOW STONES

Considering next the stones of yellow color, we have the following species to deal with: (1) diamond, (2) corundum, (3) precious topaz, (4) quartz, (5) beryl, (6) zircon, (7) tourmaline.

Yellow Zircon Resembles Yellow Diamond. Here we have less opportunity to judge of the species by the color than was the case with the red stones. The diamond, of course, is easy to tell, not by the kind of yellow that it displays, for it varies greatly in that respect, but rather by its prismatic play blended with the intrinsic color. Its luster also gives an immediate clue to its identity. It is necessary, however, to be sure that we are not being deceived by a yellow zircon, for the latter has considerable "fire" and a keen luster. Its strong double refraction and its relative softness, as well as its great density will serve to distinguish it. Of the other yellow stones, the true or precious topaz is frequently inclined to a pinkish or wine yellow and many such stones lose all their yellow (retaining their pink) when gently heated. The so-called "pinked" topazes are thus produced.

The yellow corundum rarely has a color that is at all distinctive. As far as color goes the material might be yellow quartz, or yellow beryl, or yellow zircon, or yellow tourmaline (Ceylon type). Many of the yellowish tourmalines have a decidedly greenish cast (greenish yellow chrysoberyl might resemble these also). However, in general if one has a yellow stone to determine it will be safer to make specific gravity or hardness tests, or both, before deciding, rather than to rely upon color.

LESSON XII

COLOR—*Continued*

GREEN STONES

LET us first consider what mineral species are most likely to give us green stones. Omitting the semi-precious opaque or translucent stones we have:

1. Grass-green beryl (the emerald) which is, of course, first in value among the green stones and first in the fine quality of its color.

2. Tourmaline (some specimens of which perhaps more nearly approach the emerald than any other green stones).

3. The demantoid garnet (sometimes called "olivine" in the trade).

4. True olivine (the peridot and the chrysolite of the trade).

5. Bluish green beryl (aquamarine).

6. Green sapphire (Oriental emerald or Oriental aquamarine).

7. Chrysoberyl (alexandrite and also the greenish yellow chrysoberyl).

1. Considering first the emerald, we have as legitimate a use of color in distinguishing a stone as could be selected, for emerald of fine grass-green color is not equaled by any other precious stone in the rich velvety character of its color. We have to beware here, however, of the fine glass imitations, which, while lacking the variety of true emerald, because of lack of dichroism, are nevertheless of a color so nearly like that of the emerald that no one should attempt to decide by color alone as to whether a stone is genuine or imitation emerald. If a hardness test shows that the material is a genuine hard stone and not a paste, then one who is well accustomed to the color of fine emerald can say at once whether a stone is a fine emerald or some other hard green stone. Where

the color is less fine, however, one might well refuse to decide by the color, even when sure that the material is not glass, for some fine tourmalines approach some of the poorer emeralds in richness of color.

The "Scientific Emerald" Fraud. No "scientific" emeralds of marketable size have ever been produced as far as can be learned. Many attempts to reproduce emerald by melting beryl or emerald of inferior color have resulted only in the production of a beryl glass, which, while its color might be of desirable shade, was softer and lighter in weight than true emerald. It was also a true glass and hence singly refracting and without dichroism, whereas emerald is crystalline (not glassy or amorphous), is doubly refracting, and shows dichroism.

Do not be misled, then, into buying or selling an imitation of emerald under the terms "synthetic," "scientific," or "reconstructed," as such terms, when so used, are used to deceive one into thinking that the product offered bears the same relation to the true emerald that scientific rubies and sapphires bear to the natural stones. Such is not the case.

About the most dangerous imitation of the emerald that is ever seen in the trade is the triplet that has a top and a back made of true but pale beryl (the same mineral as emerald, but not of the right color) and a thin slice of deep emerald green glass laid between. This slice of glass is usually placed behind the girdle so that a file will not find any point of attack. The specific gravity of the triplet is practically that of emerald, its color is often very good, and it is doubly refracting. It is thus a dangerous imitation. (See Fig. 8.)

Emerald Triplets. A careful examination of one of these triplets, in the unset condition, with a good lens, will reveal the thin line of junction of the beryl with the glass. (The surface lusters of the two materials are enough different for the trained eye to detect the margin at once.) Such a triplet, if held in the sun, will reflect onto a card two images in pale or white light, one coming from the top surface of the table and the other from the top surface of

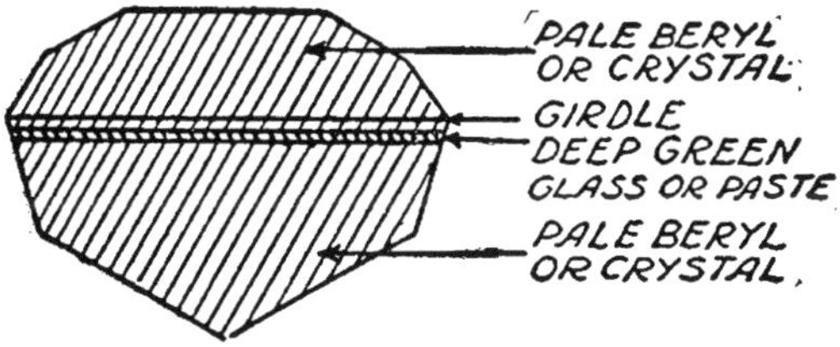

FIG. 8.—EMERALD TRIPLET.

the glass slice within. In other words, it acts in this respect like a doublet. A true emerald would give only one such reflection, which would come from the top surface of the table.

2. Tourmalines, when green, are usually darker than emeralds and of a more pronounced yellow green, or they may be of too bluish a green, as is the case with some of the finest of the green tourmalines from

Maine. Connecticut green tourmaline tends more to the dark yellowish green, and Ceylon tourmaline to the olive green. The stronger dichroism of the tourmaline frequently reveals itself to the naked eye, and there is usually one direction or position in which the color of the stone is very inferior to its color in the opposite direction or position. Most tourmalines (except the very lightest shades) must be cut so that the table of the finished stone lies on the side of the crystal, as, when cut with the table lying across the crystal (perpendicular to the principal optical axis) the stones are much too dark to be pretty. Hence when one turns the cut stone so that he is looking in the direction which was originally up and down the crystal (the direction of single refraction and of no dichroism) he gets a glimpse of a less lovely color than is furnished by the stone in other positions. With a true emerald no such disparity in the color would appear. There might be a slight change of shade (as seen by the naked eye), but no trace of an ugly shade would appear.

By studying many tourmalines and a few emeralds one may acquire an eye for the differences of color that characterize the two stones, but it is still necessary to beware of the fine glass imitation and to use the file and also to look with a high-power glass for any rounding bubbles. The emerald will never have the latter. The glass imitation frequently does have them. The sharp jagged flaws and cracks that so often appear in emerald are likely to appear also in tourmaline as both are brittle materials. The glass imitations frequently have such flaws put into them either by pinching or by striking the material. Frequently, too, wisps of tiny air bubbles are left in the glass imitations in such fashion that unless one scrutinizes them carefully with a good lens they strongly resemble the flaws in natural emerald.

I have thus gone into detail as to how one may distinguish true emerald from tourmaline and from glass imitations because, on account of the high value of fine emerald and its infrequent occurrence, there is perhaps more need for the ability to discriminate between it and its imitations and substitutes than there is in almost any other case. Where values are high the temptation to devise and to sell imitations or substitutes is great and the need for skill in distinguishing between the real and the false is proportionally great.

3. The demantoid garnet (often unfortunately and incorrectly called "olivine" in the trade) is usually of an olive or pistachio shade. It may, however, approach a pale emerald. The refraction being single in this, as in all garnets, there is little variety to the color. The dispersion being very high, however, there is a strong tendency, in spite of the depth of the body color, for this stone to display "fire," that is, rainbow color effects. The luster, too, is diamond-like as the name "demantoid" signifies. With this account of the

stone and a few chances to see the real demantoid garnet beside an emerald no one would be likely to mistake one for the other. The demantoid garnet is also very soft as compared with emerald (6½ as against nearly 8).

4. True olivine (the peridot or the chrysolite of the trade) is of a fine leaf-green or bottle-green shade in the peridot. The chrysolite of the jeweler is usually of a yellower green. Frequently an olive-green shade is seen. The luster of olivine (whether of the peridot shade or not) is oily, and this may serve to distinguish it from tourmaline (which it may resemble in color). Its double refraction is very large also, so that the doubling of the edges of the rear facets may easily be seen through the table with a lens. The dichroism is feeble too, whereas that of tourmaline is strong. No one would be likely to confuse the stone with true emerald after studying what has preceded.

5. Bluish green beryl (aquamarine) is usually of a pale transparent green or blue green (almost a pure pale blue is also found).

Having all the properties of its more valuable variety, emerald, the pale beryl may, by the use of these properties, be distinguished from the pale blue-green topaz which so strongly resembles it in color.

6. Green sapphire seldom even approaches emerald in fineness of color. When it remotely suggests emerald it is called "Oriental" emerald to denote that it is a corundum gem. Most green sapphires are of too blue a green to resemble emerald. Some are really "Oriental" aquamarines. In some cases the green of the green sapphire is due to the presence, within the cut stone, of both blue and yellow portions, the light from which, being blended by its reflection within the stone, emerges as green as seen by the unaided eye, which cannot analyze colors. The dark sapphires of Australia are frequently green when cut in one direction and

deep blue when cut in the opposite direction. The green, however, is seldom pleasing.

7. Chrysoberyl as usually seen is of a yellowish green. The fine gem chrysoberyls known as alexandrites, however, have a pleasing bluish green or deep olive green color by daylight and change in a most surprising fashion by artificial light under which they show raspberry red tints. This change, according to G. F. Herbert-Smith, is due principally to the fact that the balance in the spectrum of light transmitted by the stone is so delicate that when a light, rich in short wave lengths, falls upon it the blue green effect is evident, whereas when the light is rich in long wave lengths (red end of the spectrum), the whole stone appears red. The strong dichroism of the species also aids this contrast. The chrysoberyls of the cat's-eye type (of fibrous or tubular internal structure) are usually of olive green or brownish green shades.

Those who wish to further study color distinctions in green stones are recommended to see the chapters on beryl (pp. 184–196), peridot (pp. 225–227), corundum (pp. 172–183), tourmaline (pp. 219–224), chrysoberyl (pp. 233–237), and garnet (demantoid, pp. 216–218) in G. F. Herbert-Smith's *Gem-Stones*.

LESSON XIII

COLOR—*Continued*

BLUE STONES

THE species that furnish blue stones in sufficient number to deserve consideration are, aside from opaque stones:

1. Corundum (sapphire).
2. Spinel.
3. Tourmaline.
4. Topaz.
5. Diamond.
6. Zircon.

1. Of these minerals the only species that furnishes a fine, deep velvety blue stone is the corundum, and fine specimens of the cornflower blue variety are very much in demand and command high prices. The color

in sapphires ranges from a pale watery blue through deeper shades (often tinged with green) to the rich velvety cornflower blue that is so much in demand, and on to dark inky blues that seem almost black by artificial light. Most sapphires are better daylight stones than evening stones. Some of the sapphires from Montana, however, are of a bright electric blue that is very striking and brilliant by artificial light.

How Sapphires should be Cut. The direction in which the stone is cut helps determine the quality of the blue color, as the "ordinary" ray (sapphire exhibits dichroism) is yellowish and ugly in color, and if allowed to be conspicuous in the cut stone, its presence, blending with the blue, may give it an undesirable greenish cast. Sapphires should usually be cut so that the table of the finished stone is perpendicular to the principal optical axis of the crystal. Another way of expressing this fact is that the table should cross the long axis of the usual hexagonal crystal of sapphire, at right angles. This scheme of cutting puts the direction of single refraction up and down the finished stone, and leaves the ugly ordinary rays in poor position to emerge as the light that falls upon the girdle edges cannot enter and cross the stone to any extent.

To find out with a finished stone whether or not the lapidary has cut it properly as regards its optical properties one may use the dichroscope, and if there is little or no dichroism in evidence when looking through the table of the stone it is properly cut.

Where a sapphire shows a poor color and the dichroscope shows that the table was laid improperly, there is some possibility of improving the color by recutting to the above indicated position. However, one must use much judgment in such a case, as sapphires, like other corundum gems, frequently have their color irregularly distributed, and the skillful lapidary will place the culet of the stone in a bit of good color, and thus make the whole stone appear to better advantage. It would not do to alter such an arrangement, as one would get poorer rather than better color by recutting in such a case.

While some of the blue stones about to be described may resemble inferior sapphires, none of them approaches the better grades of sapphire in fineness of blue coloration. The scientific sapphire, of course, does approach and even equals the natural sapphire so that one must know how to distinguish between them. This distinction is not one of color, however, and it will be separately considered a little later.

2. Blue spinels are infrequently seen in commerce. They never equal the fine sapphire in their color, being more steely. They, of course, lack dichroism and are softer than sapphire as well as lighter.

3. Blue tourmalines are never of fine sapphire blue. The name indicolite which

mineralogists give to these blue stones suggests the indigo blue color which they afford. The marked dichroism of tourmaline will also help detect it. Some tourmalines from Brazil are of a lighter shade of blue and are sometimes called "Brazilian sapphires."

4. Blue topaz is usually of a pale sky blue or greenish blue and is likely to be confused with beryl of similar color. The high density of topaz (3.53) as compared with beryl (2.74) serves best to distinguish it.

"Fancy" Blue Diamonds. 5. Blue diamonds are usually of very pale bluish or violet tint. A few deeper blue stones are seen occasionally as "fancy" diamonds. These are seldom as deep blue as pale sapphires. Even the famous Hope Blue Diamond, a stone of about forty-four carats and of great value, is said to be too light in color to be considered a fine sapphire blue. Some of the deeper blue diamonds have a steely cast. The so-called blue-white stones are rarely blue in their body color, but rather are so nearly white that the blue parts of the spectra which they produce are very much in evidence, thus causing them to face up blue. There is little likelihood of mistaking a bluish diamond for any other stone on account of the "fire" and the adamantine luster of the diamond.

6. Blue zircon, however, has nearly adamantine luster and considerable fire. The color is usually sky blue. Such stones are seldom met with in the trade.

For a more detailed account of the various blue stones see G. F. Herbert-Smith's *Gem-Stones*, as follows:

For sapphires, pp. 172–173, 176, 182; for spinel, pp. 203, 204, 205; for tourmaline, pp. 220, 221, 223; for topaz, pp. 198, 200, 201; for diamond, pp. 130, 136, 170, and for zircon, pp. 229, 231.

LESSON XIV

COLOR—*Concluded*

PINK, PURPLE, BROWN, AND COLORLESS STONES

Pink Stones. Pink stones are yielded by (1) corundum (pink sapphire), (2) spinel (balas ruby), (3) tourmaline (rubellite), (4) true topaz (almost always artificially altered), (5) beryl (morganite), (6) spodumene (kunzite), and (7) quartz (rose-quartz).

These pink minerals are not easily differentiated by color alone, as the depth and quality of the pink vary greatly in different specimens of the same mineral and in the different minerals. There is dichroism in the cases of pink sapphire, pink tourmaline (strong), pink topaz (strong), pink beryl (less pronounced), and kunzite (very marked and with a yellowish tint in some directions that contrasts with the beautiful violet tint in another direction in the crystal). Pink quartz is almost always milky, and shows little dichroism. Pink spinel is without dichroism, being singly refracting. Hardness and specific gravity tests will best serve to distinguish pink stones from each other. The color alone is not a safe guide.

Purple Stones. Among the mineral species that furnish purple stones, (1) quartz is preeminent in the fineness of the purple color. Such purple stones are, of course, known as amethysts. After quartz come (2) corundum (Oriental amethyst), (3) spinel (almandine spinel), (4) garnet (almandine), and (5) spodumene (variety kunzite).

The purple of the amethyst varies from the palest tints to the full rich velvety grape-purple of the so-called Siberian amethysts. The latter are of a reddish purple (sometimes almost red) by artificial light, but of a fine violet by daylight. No other purple stone approaches them in fineness of coloring, so that here we have a real distinction based on color alone. If the purple is paler, however, one cannot be sure of the mineral by its color. Purple corundum (Oriental amethyst) is seldom as fine in color as ordinary amethyst, and never as fine as the best amethyst. It is usually of a redder purple, and by artificial light is almost ruby-like in its color.

Purple spinels are singly refracting, and lack dichroism, and hence lack variety of color. Almandine garnets also show no dichroism and lack variety of color. The garnets are, as a rule, apt to be more dense in color than the spinels.

Purple spodumene (kunzite) is pinkish to lilac in shade—usually pale, unless in large masses, and it shows very marked dichroism. A yellowish cast of color may be seen in certain directions in it also, which will aid in distinguishing it from other purple stones.

Brown Stones. (1) Diamond, (2) garnet,

(3) tourmaline, and (4) zircon furnish the principal brown stones.

Diamond, when brown, unless of a deep and pleasing color, is very undesirable, as it absorbs much light, and appears dirty by daylight and dark and sleepy by artificial light. When of a fine golden brown a diamond may have considerable value as a "fancy" stone. Such "golden fancies" can be distinguished from other brown stones (except perhaps brown zircons) by their adamantine luster, and their prismatic play or "fire."

Brown garnet (hessonite or cinnamon stone), sometimes wrongly called hyacinth in the trade, is of a deep reddish brown color. Usually the interior structure, as seen under a lens, is streaky, having a sort of mixed oil and water appearance.

Brown tourmaline is sometimes very pleasing in color. It is deep in shade, less red than cinnamon stone, and with marked dichroism, which both brown diamond and brown garnet lack

Brown zircon, while lacking dichroism, is frequently rich and pleasing in shade, and when well cut is very snappy, the luster being almost adamantine, the dispersion being large, and the refractive index high. It is useless to deny that by the unaided eye one might be deceived into thinking that a fine brown zircon was a brown diamond. However, the large double refraction of the zircon easily distinguishes it from diamond (use the sun-light-card method or look for the doubling of the edges of the rear facets as seen through the table). The relative softness (7½) also easily differentiates it from diamond.

Colorless Stones. Few colorless stones other than diamond, white sapphire(chiefly scientific), and quartz are seen in the trade. Colorless true topaz is sometimes sold and artificially whitened zircon (jargoon) is also occasionally met with. Beryl of very light green tint or even entirely colorless may also be seen at times.

Such colorless stones must of course be distinguished by properties other than color. They are mentioned here merely that the learner may be aware of what varieties of gem minerals occur in the colorless condition, and that all these minerals also occur with color in their more usual forms. This does not even except the diamond, which is rarely truly colorless.

LESSON XV

HOW TO TELL SCIENTIFIC STONES FROM NATURAL GEMS

IT should be said first that the only true scientific or synthetic stones on the market are those having the composition and properties of corundum, that is to say, the ruby and the several color varieties of sapphire, as blue, pink, yellow, and white. There is also a greenish stone that appears reddish by artificial light, which is called scientific alexandrite but which has, however, the composition and properties of the corundum gems rather than those of true alexandrite. All so-called "scientific emeralds" have proved to be either of paste of one sort or another, or else triplets having a top and a back of some inexpensive but hard stone of pale color, and a central slice of deep green glass, the three pieces being cemented together so skillfully that the junctions frequently escape any but a very careful examination with a lens.

All Scientific Stones Are Corundum Gems. Now the fact that all true scientific stones are corundum gems makes their determination fairly simple on the following basis: Among the considerable number of corundum gems of nature, whether ruby or sapphire of various colors, there is seldom found one that is entirely free from defects. Almost always, even in what are regarded as fine specimens, one will easily find with a glass, defects in the crystallization. Moreover these defects are characteristic of the corundum gems.

The scientific corundum gems, however, never have these specific defects. Hence the surest and simplest way of distinguishing between the two kinds of stones is to acquaint oneself with the typical defects of natural corundum gems, and then to look for such defects in any specimen of ruby or sapphire that is in question.

While a description of some of the most common of the typical defects of rubies and sapphires is to follow, the jeweler, who may not yet be familiar with them by actual experience, owes it to himself and to his customers to acquaint himself at first hand with the natural defects of such material, which he is always in a position to do through the courtesy of representatives of houses dealing in precious stones, if he himself does not carry such material in stock.

Typical Defects of Natural Corundum Gems. Perhaps the most common of the defects of natural corundum gems is the peculiar appearance known as "silk." This is best seen when a strong light is allowed to stream through the stone at right angles to the observer's line of sight. Sets of fine, *straight*, parallel lines will be seen, and these will frequently meet other sets of similar lines at an angle of 120 degrees (like the angle at which the sides of a regular hexagon meet) or the lines may cross each other at that angle or at an angle of 60 degrees (the supplement of 120 degrees). Such *straight* parallel lines are never seen in scientific stones, and their presence may be taken to indicate positively that the stone having them is a natural stone. In fine specimens of natural ruby or sapphire such lines will be few and difficult to find, but in some position or other they will usually be found if the search is even as careful as that which one would habitually employ in looking for defects in a diamond. In the vast majority of cases no such careful search will be required to locate "silk" in natural rubies, and if a stone that is apparently a ruby is free from such defects it is almost a foregone conclusion it is a scientific stone.

Another common type of defect in corundum gems is the occurrence of patches of milky cloudiness within the material. A little actual acquaintance with the appearance of this sort of defect in natural stones will make it easy to distinguish from the occasional cloudiness found in scientific stones, which latter cloudiness is due to the presence of swarms of minute gas bubbles. These tiny bubbles can be seen under a high power lens, and this suggests a third feature that may be used to tell whether one has a natural stone or not.

Natural rubies and sapphires, like scientific ones, frequently contain bubbles, but these are always *angular* in the natural stones, while those of the scientific stones are generally *round* or rounding, never angular.

To sum up the suggestions already presented it may be said that, since natural and scientific corundum gems are composed of essentially the same material, and have identically the same physical and chemical properties, and frequently very closely resemble each other in color, it is necessary to have recourse to some other means of distinguishing between them. The best and simplest means for those who are acquainted with the structural defects common to natural corundum gems is to seek for such defects in any specimen that is in question, and if no such defects can be found, to be very sceptical as to the naturalness of the specimen, inasmuch as perfect corundum gems are very rare in nature, and when of fine color command exceedingly high prices. No jeweler can afford to risk his reputation for knowledge and for integrity by selling as a natural stone any gem which does not possess the minor defects common to practically all corundum gems.

Structural Defects of Scientific Stones. So far our tests have been mostly negative. It was said, however, that spherical bubbles sometimes appear in scientific gems. Another characteristic *structural* defect of practically every scientific gem may be utilized to distinguish them. As is well known, the rough material is formed in boules or pear-shaped drops under an inverted blowpipe. The powdered material is fed in with one of the gases and passes through the flame, melting as it goes, and then accumulating and crystallizing below as a boule. The top or head of this boule is rounding from the start, and hence the successive layers of material gather in thin curved zones. The color and structure of these successive zones are not perfectly uniform, hence when cut stones are made from the boules these *curving* parallel layers may be seen within by the use of a good lens, especially if the cut stone is held in a strong crossing light, as was suggested when directions were given above as to the best way to look for "silk" in a natural stone.

Owing to the shape of a well cut stone it is sometimes difficult to get light through the material, yet by turning the stone repeatedly, some position will be found in which the curving parallel striæ can be seen. They are easily seen in scientific ruby, less easily in dark blue sapphire, but still they can be found on close search. In the light colored stones and in white sapphire, the difficulty is greater, as there are no color variations in the latter case. However, the value of white sapphire is so slight, whether natural or artificial, that it is a matter of but little moment, and what has already been said as to natural defects, applies to white sapphire as well as to the colored varieties, and absolutely clear and perfect natural white sapphire is rare.

One more distinguishing mark of the scientific stones may be added to give full measure to the scheme of separation, that no one need be deceived.

The surface finish of the scientific stones is rarely as good as that of the natural material and it appears to be more difficult to produce a good polish on scientific stones than on natural ones. The degree of hardness of the scientific stones seems to be slightly variable in different parts of the same piece so that the polishing material removes the surface material unequally, leaving minute streaky marks on the surfaces of the facets. Possibly this condition might be remedied by skillful treatment, but hardly at the price obtainable for the product, so that a close study of the surface finish will sometimes help in distinguishing between natural and artificial material. Any fine specimen of natural ruby or sapphire will have usually received very expert treatment and a splendid surface finish.

In conclusion, then, the points to be remembered in determining the origin of corundum gems are four in number.

1. Expect to find natural defects, such as "silk" or cloudy patches, or *angular* bubbles in all natural stones.

2. If bubbles are present in artificial material they will be *round* or rounding.

3. Artificial material will always have *curving* parallel striæ within it.

4. The *surface finish* of artificial material is seldom or never equal to that of natural material.

It ought not to be necessary to add that material from either source may be cut to any shape, and that artificial rubies may be seen in most Oriental garb, hence all specimens should have applied to them the above tests regardless of the seeming antiquity of their cut or of their alleged pedigree.

LESSON XVI

HOW TO TEST AN "UNKNOWN" GEM

HAVING now considered separately the principal physical properties by means of which one can identify a precious stone, let us attempt to give as good an idea as the printed page can convey of how one should go about determining to what species a gem belongs.

Signs of Wear in an Emerald. To make the matter more concrete, and therefore more interesting, let us consider a real case, the most recent problem, in fact, that the author has had to solve. A lady of some wealth had purchased, for a large sum, a green stone which purported to be an emerald. After a few years of wear as a ring stone she noticed one day that the stone had dulled around the edges of its table, and thinking that that ought not to be the case with a real emerald, she appealed to a dealer in diamonds to know if her stone was a real emerald. The diamond merchant told her frankly that, while he was competent in all matters pertaining to diamonds, he could not be sure of himself regarding colored stones, and advised the lady to see the author.

The matter being thus introduced, the lady was at once informed that even a real emerald might show signs of wear after a few years of the hard use that comes to a ring stone.

While emerald has, as we saw in the lesson on hardness, a degree of hardness rated as nearly 8 (7½ in the table), it is nevertheless a rather brittle material and the long series of tiny blows that a ring stone is bound to meet with will cause minute yielding along the exposed edges and corners of the top facets. This being announced, the first step in the examination of the stone was to clean it and to give it a careful examination with a ten-power lens. (An aplanatic triplet will be found best for this purpose.)

Color. The color was, of course, the most obvious property, but, as has already been said, color is not to be relied upon in all cases. In this case the color was a good emerald green but a bit bluer than the finest grass green. A very fine Maine tourmaline might approach this stone in color, so it became necessary to consider this possibility. A glass imitation, too, might have a color equal or superior to this.

Imperfections. While noting the color, the imperfections of the stone claimed attention. They consisted mainly of minute jagged cracks of the character peculiar to brittle materials such as both emerald and tourmaline. So far it will be noted either of the above minerals might have furnished the lady's gem. As glass can be artificially crackled to produce similar flaws the stone might have been only an imitation as far as anything yet learned about it goes.

File Test. The next step was to test its hardness by gently applying a very fine file to an exposed point at one corner of the girdle. The file slipped on the material as a skate slips on ice. Evidently we did not have to do with a glass imitation.

Refraction. Knowing now that we had a true hard mineral, it remained to be determined what mineral it was. On holding the stone in direct sunlight and reflecting the light onto a white card it was seen at once that the material was doubly refracting, for a series of *double* images of the back facets appeared. These double images might have been produced by tourmaline as well as by emerald. (Not however by glass which is singly refracting.) If a direct reading refractometer had been available the matter could have been settled at once by reading the refractive indices of the material, for tourmaline and emerald have not only different refractive indices but have double refraction to different degrees. Such an instrument was not available at the time and will hardly be available to most of those who are studying this lesson, so we can go on with our account of the further testing of the green stone.

Hardness. A test upon the surface of a quartz crystal showed that the stone was harder than quartz (but so is tourmaline). A true topaz crystal was too hard for the ring stone, whose edge slipped over the smooth topaz surface. The green stone was therefore not a green corundum (Oriental emerald) as the latter has hardness 9 and scratches topaz.

With hardness evidently between 7 and 8 and with double refraction and with the kind of flaws peculiar to rather brittle minerals we had in all probability either a tourmaline or an emerald.

Dichroism. The dichroscope (which might have been used much earlier in the test but was not at hand at the time) was next tried and the stone was seen to have marked dichroism—a bluish green and a yellowish green appearing in the two squares of the instrument when the stone was held in front of the opening and viewed against a strong light.

As either tourmaline or emerald might thus exhibit dichroism (the tourmaline more strongly, however, than the emerald) one more test was tried to finally decide the matter.

Specific Gravity. The stone was removed from its setting and two specific gravity determinations made by means of a specific gravity bottle and a fine chemical balance. The two results, which came closely alike, averaged 2.70 which agrees very nearly with emerald (2.74) and which is far removed from the specific gravity of tourmaline (3.10). The stone was now *definitely known* to be an emerald, as each of several tests agreed with the properties of emerald, namely:

Color—nearly grass green.

Imperfections—like those of emerald.

Hardness—7½.

Refraction—double.

Dichroism—easily noted.

Specific gravity—2.70.

While one who was accustomed to deal in fine emeralds might not need to make as detailed an examination of the stone as has just been indicated above, yet for most of us who do not have many opportunities of studying valuable emeralds it is safer to make sure by complete tests.

One other concrete example of how to go about testing unknown stones must suffice to conclude this lesson, after which the student, who has mastered the separate lessons preceding this, should proceed to test as many "unknowns" as his time and industry permit in order to really make *his own* the matter of these lessons. It may be added here that the task of testing a stone is much more rapid than this laborious effort to teach others how to do it might indicate. To one skilled in these matters only a few seconds are required for the inspection of a stone with the lens, the dichroscope, or the refractometer, and hardness tests are swiftly made. A specific gravity test requires more time and should be resorted to only when there remains a reasonable doubt after other tests have been applied.

Now for our final example. A red stone, cut in the form of a pear shaped brilliant, was submitted to the writer for determination. It had been acquired by an American gentleman in Japan from an East Indian who was in financial straits. Along with it, as security for a loan, the American obtained a number of smaller red stones, a bluish stone, and a larger red stone. The red stones were all supposed to be rubies. On examination of the larger red stone with a lens it was at once noted that the internal structure was that of *scientific ruby*.

Testing Other Stones. Somewhat dashed by the announcement of this discovery the owner began to fear that all his gems were false. Examination of the small red stones showed abundance of "silk," a peculiar fibrous appearance within the stone caused by its internal structure. The fibers were *straight* and *parallel*, not *curved* and *parallel* as in synthetic ruby. Tiny bubbles of angular shape also indicated that the small stones were natural rubies. They exhibited dichroism and scratched topaz and it was therefore decided that they at least were genuine.

The pear shaped brilliant which was first mentioned was of a peculiar, slightly yellowish, red color. It was very pellucid and free from any striæ either of the straight or curved types. It had in fact no flaws except a rather large nick on one of the back surfaces near the girdle. This was not in evidence from the front of the stone and had evidently been left by the Oriental gem cutter to avoid loss in weight while cutting the stone.

The peculiar yellowish character of the red color led us to suspect ruby spinel. The stone was therefore inspected with the dichroscope and found to possess no dichroism. The sunlight-card test, too, showed that the stone was singly refracting.

A test of the hardness showed that the material barely scratched topaz, but was attacked by sapphire. It was therefore judged to be a red spinel.

The large bluish stone which the gentleman acquired with the red stones proved to be iolite, sometimes called cordierite or water-sapphire (*Saphire d'eau*), a stone seldom seen in this country. It had marked dichroism—showing a smoky blue color in one direction and a yellowish white in another. The difference was so marked as to be easily seen without the dichroscope.

LESSON XVII

SUITABILITY OF STONES FOR VARIOUS TYPES OF JEWELS AS DETERMINED BY HARDNESS, BRITTLENESS, AND CLEAVABILITY

Hard Stones not Necessarily Tough. As was suggested in the lesson on hardness there is prevalent in the public mind an erroneous belief that hardness carries with it ability to resist blows as well as abrasion. Now that *it does not follow that because a precious stone is very hard, it will wear well*, should be made plain. Some rather hard minerals are seldom or never used as gems, in spite of considerable beauty and hardness, because of their great brittleness. Other stones, while fairly hard and reasonably tough in certain directions, have nevertheless so pronounced a cleavage that they do not wear well if cut, and are sometimes very difficult to cut at all.

In view of these facts it will be well to consider briefly what stones, among those most in use, are sufficiently tough as well as hard, to give good service in jewels, such as rings, which are subject to rough wear. We may also consider those stones, whose softness, or brittleness, or ready cleavability, requires that they should be reserved for use only in those jewels which, because of their nature, receive less rough usage.

In order to deal with the principal gems systematically, let us consider them in the order of their hardness, beginning with the hardest gem material known, which is, of course, diamond.

Durability of the Diamond. Fortunately this king of gems possesses in addition to its great hardness, considerable toughness, and although it is readily cleavable in certain directions it nevertheless requires a notable amount of force applied in a particular direction to cause it to cleave. Although sharp knocks will occasionally flake off thin layers from diamonds when roughly worn in rings, or even in extreme cases fracture them, yet this happens but seldom and, as the enormous use of the diamond in ring mountings proves, it is entirely suitable for that purpose. It follows that, if a stone can stand ring usage, it can safely be used for any purpose for which precious stones are mounted.

The Corundum Gems. Next after the diamond in hardness come the corundum gems, *i. e.*, ruby, sapphire, and the series of corundum gems of colors other than red and blue. These stones have no noticeable cleavage and are exceedingly tough, for minerals, as well as very hard. We have only to consider the use of impure corundum (emery) as a commercial abrasive in emery wheels, emery cloth, emery paper, etc., to see that the material is tough. Any of the corundum gems therefore may be used in any type of jewel without undue risk of wear or breakage. Customers of jewelers should, however, be cautioned against wearing ruby or sapphire rings on the same finger with a diamond ring in cases where it would be possible for the two stones to rub against each other. So much harder than the ruby is the diamond (in spite of the seeming closeness of position in Mohs's scale) that the slightest touch upon a ruby surface with a diamond will produce a pronounced scratch. The possessor of diamonds and other stones should also be cautioned against keeping them loose in the same jewel case or other container, as the shaking together may result in the scratching of the softer materials. The Arabs are said to have a legend to the effect that the diamond is an *angry* stone and that it should not be allowed to associate with other stones lest it scratch them.

Chrysoberyl. Passing on to the next mineral in the scale of hardness we come to chrysoberyl, which is rated as 8½ on Mohs's scale. This mineral furnishes us the gem, alexandrite, which is notable for its power to change in color from green in daylight to red in artificial light. Chrysoberyl also supplies the finest cat's-eyes (when the material is of a sufficiently fibrous or tubular structure), and it further supplies the greenish yellow stones frequently (though incorrectly) called "chrysolite" by jewelers. The material is very hard and reasonably tough and may be used in almost any suitable mounting.

Spinel. After chrysoberyl come the materials rated as about 8 in hardness. First and hardest of these is spinel, then comes true or precious topaz. The various spinels are very hard and tough stones. The rough material persists in turbulent mountain streams where weaker minerals are ground to powder, and when cut and polished, spinel will wear well in any jewel. The author has long worn a ruby spinel in a ring on the right hand and has done many things that have subjected it to hard knocks, yet it is still intact, except for a spot that accidentally came in contact with a fast-flying carborundum wheel, which of course abraded the spinel.

Topaz. The true topaz is a bit softer than spinel, and the rough crystals show a very perfect basal cleavage. That is, they will cleave in a plane parallel to the bases of the usual orthorhombic crystals. This being the case a cut topaz is very likely to be damaged by a blow or even by being dropped on a hard surface, and it would be wiser not to set such a stone in a ring unless it was to be but little used, or used by one who would not engage in rough work while wearing it. Thus a lady might wear a topaz ring on dress occasions for a long time without damaging it, but it would not do for a machinist to wear one in a ring.

Gems between 7 and 8 in Hardness. We now come to a rather long list of gem minerals ranging between 7 and 8 in hardness. Of these the principal ones are zircon, almandine garnet, and beryl (emerald and aquamarine) rated as 7½ in hardness, and pyrope and hessonite garnet rated as $7\frac{1}{4}$ in hardness. Tourmaline and kunzite may also be included in this group as being on the average slightly above 7 in hardness.

The above minerals are all harder than quartz, and hence not subject to abrasion by the quartz dust which is everywhere present. In this respect they are suitable for fairly hard wear. The garnets are of sufficient toughness so that they may be freely used in rings—and the extensive use of thin slices of garnet to top doublets proves the suitability of the material for resisting wear. The zircon is rather more brittle and the artificially whitened zircons (known as jargoons) are especially subject to breakage when worn in rings. Fortunately jargoons are not commonly sold.

The beryl, whether emerald or aquamarine, is rather brittle. Emeralds are seldom found in river gravels. The material cannot persist in the mountain streams that bring down other and tougher minerals. The extreme beauty and value of the emerald has led to its use in the finest jewels, and the temptation is strong to set it in rings, especially in rings for ladies. If such rings are worn with the care that valuable jewels should receive they will probably last a long time without any more serious damage than the dulling of the sharp edges of the facets around the table. This slight damage can at any time be repaired by a light repolishing of the affected facets. If an emerald is already badly shattered, or as it is called "mossy" in character, it will not be wise to set it in a ring, as a slight shock might complete its fracture. What has been said about emerald applies equally to aquamarine except that the value at stake is much less and the material is usually much freer from cracks.

Tourmalines, like emeralds, are brittle, and should be treated accordingly. Here, however, we are dealing with a much less expensive material than emerald, and if a customer desires a tourmaline in a ring mounting, while it will be best to suggest care in wearing it, the loss, in case of breakage, will usually be slight.

Kunzite, like all spodumene, has a pronounced cleavage. It should therefore be used in brooches, pendants, and such jewels, rather than in rings. Lapidaries dislike to cut it under some conditions because of its fragility.

Quartz Gems. Coming down to hardness 7 we have the various quartz gems and jade (variety jadeite). The principal quartz gems are, of course, amethyst and citrine quartz (the stone that is almost universally called topaz in the trade). As crystalline quartz is fairly tough and lacks any pronounced cleavage, and as it is as hard as anything it is likely to meet with in use, it is a durable stone in rings or in other mountings. In the course of time the sharp edges will wear dull from friction with objects carrying common dust, which is largely composed of powdered quartz itself, and which therefore gradually dulls a quartz gem. Old amethysts or "topazes" that have been long in use in rings show this dulling. There is, however, little danger of fracture with amethyst or "topaz" unless the blow is severe and then any stone might yield.

The many semi-precious stones which have a quartz basis (such as the varieties of waxy or cryptocrystalline chalcedony which is largely quartz in a very minutely crystalline condition) are often even tougher than the clear crystallized quartz. Carnelian, agate, quartz cat's-eye, jasper (containing earthy impurities), and those materials in which quartz has more or less completely replaced other substances, such as silicified crocidolite, petrified wood, chrysocolla quartz, etc., are all nearly as hard and quite as tough as quartz itself, and they make admirable stones for inexpensive rings of the arts and crafts type.

Jade. Jade, of the jadeite variety, which is rarer than the nephrite jade, and more highly regarded by the Chinese, is an exceedingly tough material. One can beat a chunk of the rough material with a hammer without making much impression upon it. It is also fairly hard, about as hard as quartz, and with the two properties of toughness and hardness it possesses excellent wearing qualities in any kind of mounting. True jade, whether jadeite or nephrite, deserves a larger use in inexpensive ornaments, as it may be had of very fine green color and it is inexpensive and durable.

Softer Stones. Coming next to those minerals whose hardness is 6 or over, but less than 7, we have to consider jade of the nephrite variety, demantoid garnet ("olivine" of the trade), peridot (or chrysolite, or the olivine of the mineralogist), turquoise, moonstone, and opal.

As has already been said of jadeite, the jade of the nephrite variety, while slightly less hard, is about as tough a mineral as one could expect to find. It can take care of itself in any situation.

The demantoid garnet (the "olivine" of the trade) is so beautiful and brilliant a stone that it is a pity that it is so lacking in hardness. It will do very well for mounting in such jewels as scarf pins, lavallières, etc., where but little hard wear is met with, but it cannot be recommended for hard ring use.

The peridot, too, is rather soft for ring use and will last much better in scarf pins or other mountings little subject to rubbing or to shocks.

Turquoise, although rather soft, is fairly tough, as its waxy luster might make one suppose, and in addition, being an opaque stone, slight dulling or scratching hardly lessens its beauty. It may therefore be used in ring mountings. However, it should be suggested that most turquoise is sufficiently porous to absorb grease, oil, or other liquids, and its color is frequently ruined thereby. Of course, such a change is far more likely to occur to a ring stone than to a turquoise mounted in some more protected situation.

The moonstone, being a variety of feldspar, has the pronounced cleavage of that mineral and will not stand blows without exhibiting this property. Moonstones are therefore better suited to the less rude service in brooch mountings, etc., than to that of ring stones. However, being comparatively inexpensive, many moonstones, especially of the choicer bluish type, are set in ring mountings. The lack of hardness may be expected to dull their surfaces in time even though no shock starts a cleavage.

The Opal. There remains the opal, of hardness 6, to be considered. As is well known opal is a solidified jelly of siliceous composition, containing also combined water. It is not only soft but very brittle and it will crack very easily. Many opals crack in the paper in which they are sold, perhaps because of unequal expansion or contraction, due to heat or cold. In spite of this fragility, thousands of fine opals, and a host of commoner ones, are set in rings, where many of them subsequently come to a violent end, and all, sooner or later, become dulled and require repolishing.

The great beauty of the opal, rivaling any mineral in its color-play, causes us to chance the risk of damage in order to mount it where its vivid hues may be advantageously viewed by the wearer as well as by others.

Very Soft Stones. Of stones softer than 6 we have but few and none of them is really fit for hard service. Lapis lazuli, 5½ in hardness, has a beautiful blue color, frequently flecked with white or with bits of fool's gold. Its surface soon becomes dulled by hard wear.

Two more of the softer materials, malachite and azurite, remain to be described. These are both varieties of copper carbonate with combined water, the azurite having less water. Both take a good polish, but fail to retain it in use, being only of hardness 3½ to 4.

LESSON XVIII

MINERAL SPECIES TO WHICH THE VARIOUS GEMS BELONG AND THE CHEMICAL COMPOSITION THEREOF

ALTHOUGH we have a very large number of different kinds of precious and semi-precious stones, to judge by the long list of names to be found in books on gems, yet all these stones can be rather simply classified on the basis of their chemical composition, into one or another of a comparatively small number of mineral species. While jewelers seldom make use of a knowledge of the chemistry of the precious stones in identifying them, nevertheless such a knowledge is useful, both by way of information, and because it leads to a better and clearer understanding of the many similarities among stones whose color might lead one to regard them as dissimilar.

Mineral Species. We must first consider what is meant by a "mineral species" and find out what relation exists between that subject and chemical composition. Now by a "mineral species" is understood a single substance, having (except for mechanically admixed impurities) practically a constant chemical composition, and having practically identical physical properties in all specimens of it.

Diamond and Corundum. A chemist would call a true mineral a *pure substance*, just as sugar and salt are pure substances to the chemist. Thus *diamond* is a "mineral species," as is also *corundum*. There are many different colors of both diamond and corundum, but these different colors are believed to be due to the presence in the pure substance of impurities in small amounts. Thus every diamond consists mainly of pure carbon, and all the corundum gems (*ruby* and the various colors of *sapphire*) consist mainly of pure oxide of aluminum. The properties of all diamonds are practically alike and so are the properties of all the corundum gems whether red (ruby), blue (sapphire), yellow (Oriental topaz), green (Oriental emerald), or purple (Oriental amethyst).

Thus all diamonds, of whatever, color belong to the one species, diamond, and in this case the usual custom in naming them agrees with the facts. Similarly all sapphires, of whatever color, belong to the mineral species "corundum." Thus a ruby is a red corundum.

The old French traveler and gem merchant, Tavernier, tells us that in the seventeenth century, when he visited the mines of Pegu, the natives knew of the similarity of the corundum gems and even called all by one name, with other names attached to designate the color. Singularly enough, the common name used by them was *ruby* rather than sapphire, as now. Thus they called blue corundum gems blue rubies; yellow corundums, yellow rubies, etc.

It is easily seen that if one recognizes the similar nature of all the many colors and shades of corundum that the number of things that one has to remember in order to be well acquainted with these stones is considerably diminished. Thus, instead of having a whole series of specific gravities to remember one has only to remember that all the corundum gems have a specific gravity of approximately 4. Similarly they are all of practically the same refractive index (1.761–1.770, being doubly refracting) that they all exhibit dichroism when at all deeply colored, etc.

Having thus indicated what we mean by mineral species and having illustrated the matter by the cases of diamond and corundum and further having stated that all diamonds are composed of pure carbon (except for traces of impurities) and all corundum gems mainly of oxide of aluminum, we may proceed to consider other mineral species and find out what gems they afford us.

Carbon, the only Element Furnishing a Gem. It will be noted that the first species considered, diamond, consisted of but a single element, carbon. It is thus exceedingly simple in composition, being not only a pure substance but, in addition, an elementary substance. **Corundum,** the second species considered, was a little more complex, having two elements, aluminum and oxygen, in its make-up, but completely and definitely combined in a new compound that resembles neither aluminum nor oxygen. It is thus a compound substance. No other element than carbon affords any gem-stone when by itself.

Oxides of Metals. There is, however, another oxide, in addition to aluminum oxide, that furnishes gem material. It is *silicon oxide*, containing the two elements silicon and oxygen. Silicon itself is a dark, gray, crystalline element that seems half metallic, half non-metallic in its properties. It is never found by itself in nature but about twenty-eight per cent. of the crust of the earth is composed of it in compound forms, and one of the most abundant of these is **quartz,** which is a mineral species, and which contains just silicon and oxygen. That is, it is oxide of silicon. Now quartz is colorless when pure (*rock crystal*), but it is frequently found colored purple (probably by oxide of manganese) and it is then called *amethyst* by the jeweler. At other times its color is yellow (due to oxide of iron) and then the jeweler is prone to call it "*topaz*," although properly speaking that name should, as we shall soon see, be reserved for an entirely different mineral species. *Chalcedony* too (which when banded furnishes us our *agates*, and when reddish our *carnelian*) is a variety of quartz, and *prase* is only quartz colored green by fibers of actinolite within it.

The common *cat's-eye* and the *tiger's-eye* are varieties of quartz enclosing fibrous minerals or replacing them while still keeping the arrangement that they had. "*Venus hair stone*" is quartz containing needle-like crystals of rutile, and "*iris*" is quartz that has been crackled within, so as to produce rainbow colors, because of the effects of thin layers of material. *Aventurine quartz* (sometimes called goldstone) has spangles of mica or of some other mineral enclosed in it. The *jaspers* are mainly quartz with more of earthy impurity than the preceding stones.

Thus all this long list of stones of differing names can be classified under the one mineral species, quartz. Together they constitute the quartz gems. In properties they are essentially alike, having specific gravity 2.66, hardness 7, slight double refraction, etc., the slight differences that exist being due only to the presence of varying amounts of foreign matter.

Opal. The *opal* may be considered along with the quartz gems, because, like them, it is composed mainly of oxide of silicon, but the opal also has water combined with the silicon oxide (not merely imprisoned in it). Thus opal is a hydrous form of silica (hydrous comes from the Greek word for water).

Spinel. All our other stones are of more complicated chemical composition than the preceding. Coming now to mineral species which have three chemical elements in them we may consider first *spinel*, which has the two metallic elements aluminum and magnesium and the non-metallic element oxygen in it. It is virtually a compound of the two oxides, aluminum oxide and magnesium oxide. The variously colored spinels, like the various corundums, all have the same properties, thus they are all of hardness 8 or a little higher, they all have single refraction, and all have specific gravity 3.60.

Chrysoberyl. Another mineral species which, like spinel, has just three elements in its composition is *chrysoberyl.* This mineral contains the metals aluminum and berylium combined with the non-metal oxygen. Thus it is really to be regarded as a compound of the two oxides, aluminum oxide and berylium oxide. This species furnishes us *Alexandrite*, *chrysoberyl* *cat's-eye* and less valuable chrysoberyls of yellowish green color. All are of the one species, the marked color difference being due to the presence of different impurities. The cat's-eye effect in one of the varieties is due to the internal structure rather than to the nature of the material.

The Silicates. Nearly all of the remaining precious stones belong to a great group of mineral species known as the silicates. These are so called because they consist largely of oxide of silicon (the material above referred to under quartz gems). This oxide of silicon is not free and separate in the silicates but is combined chemically with other oxides, chiefly with metallic oxides. Thus there are many different silicates because, in the earth, many different metallic oxides have combined with silicon oxide. Also in many cases two or three or even more metallic oxides have combined with silicon oxide to make single new compounds.

Glass a Mixture of Silicates. Those who are

familiar with glass making may receive some help at this point by remembering that the various glasses are silicates, for they are made by melting sand (which is nearly pure oxide of silicon) with various metallic oxides. With lime (calcium oxide) and soda (which yields sodium oxide) we get soda-lime glass (common window glass). Lead oxide being added to the mixture a dense, very brilliant, but soft glass (flint glass) results. Cut glass dishes and "paste" gems are made of this flint glass. Now the glasses, although they are silicates, are not crystalline, but rather they are *amorphous*, that is, without any definite structure. Nature's silicates, on the other hand, are usually crystallized or at least crystalline in structure. (In a few cases we find true glasses, volcanic glass, or obsidian, for example.)

Having thus introduced the silicates we may now consider which ones among the many mineral silicates furnish us with precious or semi-precious stones.

Beryl, Emerald, and Aquamarine. First in value among the silicates is *beryl*, which, when grass green, we call *emerald*. The *aquamarine* and *golden beryl* too belong to this same species. Beryl is a silicate of aluminum and berylium. That is, it is a compound in which oxide of silicon is united with the oxides of aluminum and of berylium. There are thus four chemical elements combined in the one substance and it is hence more complicated in its composition than any of the gems that we have yet considered. It is worthy of note that aluminum occurs in the majority of precious stones, the only species so far considered that lack it being diamond, and the quartz gems.

Perhaps the silicates that are next in importance to the jeweler, after beryl, are those which form the *garnets* of various types. There are four principal varieties of garnet (although specimens of garnet frequently show a crossing or blending of the types).

Garnets. The types are (1) *Almandite* garnet; (2) *Pyrope* garnet; (3) *Hessonite* garnet; and (4) *Andradite* garnet. These are all silicates, the almandite garnets being silicates of iron and aluminum; the pyrope garnets are silicates of magnesium and aluminum; the hessonite garnets, silicates of calcium and aluminum, and the andradite garnets, silicates of calcium and iron.

The so-called almandine garnets of the jeweler are frequently of the almandite class and tend to purplish red. The pyrope garnets are, as the name literally implies, of fire red color, as a rule, but they also may be purplish in color. The hessonite garnets are frequently brownish red and are sometimes called "cinnamon stones." The andradite garnets furnish the brilliant, nearly emerald green demantoids (so often called "*olivine*" by the trade).

Thus all the garnets are silicates and yet we have these four principal mineral species, which, however, are more closely related to each other in crystal form, in character of composition and in general properties, than is usual among the other silicates. Specimens which have any one of the four types of composition unblended with any of the other types would be found to be exactly alike in properties. As was suggested above, however, there is a great tendency to blend and this is well illustrated by the magnificent *rhodolite* garnets, of rhododendron hue which were found in Macon County, North Carolina. These had a composition between almandite and pyrope, that is, they had both magnesium and iron with aluminum and silica.

The true **topaz** next calls for consideration as it too is a silicate. The metallic part consists of aluminum, and there are present also the non-metals fluorine and hydrogen. Here we have five elements in the one substance. Various specimens of this species may be wine yellow, light blue, or bluish green, pink or colorless, yet they all have essentially the same properties.

Tourmaline is about as complicated a mineral as we have. It is a very complex silicate, containing aluminum, magnesium, sodium (or other alkali metal, as, for example, lithium), iron, boron, and hydrogen. As Ruskin says of it in his *The Ethics of the Dust*, when Mary asks "and what is it made of?" "A little of everything; there's always flint (silica) and clay (alumina) and magnesia in it and the black is iron, according to its fancy; and there's boracic acid, if you know what that is: and if you don't, I cannot tell you to-day and it doesn't signify; and there's potash and soda; and on the whole, the chemistry of it is more like a mediæval doctor's prescription, than the making of a respectable mineral." The various tourmalines very closely resemble each other in their properties, the slight differences corresponding to differences in composition do not alter the general nature of the material.

Moonstone belongs to a species of mineral known as feldspar. The particular feldspar that furnishes most of the moonstone is orthoclase, a silicate of potassium and aluminum. Another feldspar sometimes seen as a semi-precious stone is *Labradorite*. *Amazonite*, also is a feldspar. *Sunstone* is a feldspar which includes tiny flakes or spangles of some other mineral.

The mineral species *olivine* gives us *peridot*. It is a silicate of magnesium.

Zircon is itself a species of mineral and is a silicate of zirconium. The names *hyacinth*, *jacinth*, and *jargoon* are applied to red, yellow, and colorless zircon in the order as given.

Jade may be of any of several different species of minerals, all of which are very tough. The principal jades belong, however, to one or the other of two species, *jadeite* and *nephrite*. Jadeite is a sodium aluminum silicate and nephrite, a calcium magnesium silicate.

Leaving the silicates we find very few gem minerals remaining. The phosphates furnish us *turquoise*, a hydrous aluminum phosphate, with copper and iron. *Varicite* is also a phosphate (a hydrated aluminum phosphate).

The carbonates give us *malachite* and *azurite*, both carbonates of copper with combined water, the malachite having more water.

LESSON XIX

THE NAMING OF PRECIOUS STONES

OWING to the confusion which may result from a lack of uniformity in the naming of precious stones, it is very desirable that jewelers and stone merchants inform themselves in regard to the correct use of the names of the gems, and that they use care in speaking and in writing such names.

As nearly all precious and semi-precious stones are derived from a relatively small number of *mineral species*, as we saw in Lesson XVIII., and as the science of *mineralogy* has a very orderly and systematic method of naming the minerals, the best results are had in the naming of gems when we use, as far as is possible, the language of mineralogy.

Ancient Usage. Long established custom

and usage, however, must be observed, for any system of naming must be generally understood in order to be useful. Thus the proper name for blood red, crystallized oxide of aluminum, of gem quality, according to the mineralogical system of naming, would be red *corundum*, but that same material is referred to in the Old Testament thus (in speaking of wisdom), "She is more precious than *rubies*." It is obviously necessary to keep and to use all such terms as have been for years established in usage, even though they do not agree with the scientific method of naming the particular mineral. It is, however, necessary that any name, thus retained, should be correctly used, and that it should not be applied to more than one material. Thus the term *ruby* should be reserved exclusively for red corundum, and not applied to other red minerals such as garnet, spinel, etc., as is too often done.

It will be the purpose of this lesson to attempt to set forth as clearly and as briefly as possible what constitutes good usage in the naming of the principal stones, and also to point out what incorrect usage is most in need of being avoided.

To cover the subject systematically we will adopt the order of hardness that we did in discussing mineral species in Lesson XVIII.

Fancy Diamonds. Beginning with the hardest of all gems, the *diamond*, we have no difficulty as regards naming, as all specimens of this mineral, regardless of color, are called diamonds. When it is necessary to designate particular colors or tints, or differences in tint, additional names are used—for example, all diamonds of pronounced and pleasing color are called "fancy" diamonds in the trade. Certain of these "fancy" diamonds are still further defined by using a name specifying the color, as, for example, "canary" diamonds (when of a fine bright yellow), or "golden fancies," when of a fine golden brown, or "orange," or "pink," or "absinthe green," or "violet," as the case may be.

Names of Various Grades of White Diamonds. The great majority of the diamonds which come on the market as cut stones belong, however, to the group which would be spoken of as white diamonds, but many qualifying names are needed to express the degree of approach to pure white possessed by different grades of these diamonds. Thus the terms: 1, *Jägers;* 2, *Rivers;* 3, *Blue Wesseltons;* 4, *Wesseltons;* 5, *Top Crystals;* 6, *Crystals;* 7, *very light brown;* 8, *Top Silver Capes;* 9, *Silver Capes;* 10, *Capes;* 11, *Yellows*, and 12, *Browns*, describe *increasing* depth of color, and hence *decreasing* value in diamonds.

Popular Names. Certain more popular names for diamonds of differing degrees of whiteness may next be set forth. The term "blue white" (a much abused expression, by the way) should be applied only to diamonds of such a close approach to pure whiteness of body substance, as seen on edge in the paper that, when faced up and undimmed, they give such a strong play of *prismatic* blue that any slight trace of yellow in their substance is completely disguised, and the effect upon the eye is notably blue. This would be the case with stones of the grades from 1 through 4 in the list above. Grades 5 and 6 might properly be called "*fine white*," and grades 7, 8, and 9 simply "*white*." Grade 10 is frequently spoken of as "*commercial white*," and grade 11 sometimes as "off color." Grade 12 includes all degrees of brownness except the very light shades and the deep, pretty shades of the "fancy" browns.

Rubies. Leaving the naming of the different colors of diamonds we come to the gems furnished us by the mineral known as *corundum*. As we have previously seen, this mineral occurs in many different colors and with wide differences of tint and shade in each of the principal colors. The best practice with regard to naming the corundum gems is to call the red material, when of a good, full red of pleasing shade, *ruby*. The finest shades of blood red are usually called "*Burmah rubies*" because more rubies of this quality are found in Burmah than anywhere else. Any ruby of the required shade would, however, be called a Burmah ruby in the trade regardless of its geographical origin. The most desirable tint among Burmah rubies is that which is known as "pigeon blood" in color. This color is perhaps more accurately defined as like the color in the center of the red of the solar spectrum. Certain slightly deeper red rubies are said to be of "beef blood" color. The English are said to prefer these. Those of slightly lighter tint than pigeon blood are sometimes referred to as of "French color," from the fact that they are preferred by French connoisseurs.

Rubies of dark, garnet-like shade are known as "*Siam rubies*," many such being found in that country. Light pinkish rubies are called

"*Ceylon rubies.*" It should be clearly kept in mind that all these "rubies" are of red corundum, and that in all their distinctive properties except color they are essentially similar.

Sapphires. Corundum of fine blue color is known as "*sapphire.*" The "cornflower blue" seems to be most in favor at present. Such sapphires are sometimes called "*Kashmir sapphires*" because many fine ones come from that State. "*Ceylon sapphires*" are usually paler than thc cornflower blue. "*Montana sapphires*" are usually of greenish blue or pale electric blue. Such fine blue stones as are mined in Montana would be sold under another name according to the quality of their color, and not as "Montana sapphires." "*Australian sapphires*" are of a very deep, inky blue, and do not command a high price. Here again, as with rubies, the classification depends upon the color rather than upon the origin, although the geographical names that are used, correctly state the usual source of stones of the particular color.

All corundums other than ruby and blue sapphire are usually called by the term "sapphire," with a qualifying adjective designating the color; thus we may have pink sapphire, golden sapphire, green sapphire, etc. When of very fine yellow color the yellow sapphire is sometimes called "*Oriental topaz*" by jewelers, the term "*Oriental*" as thus used indicating that the material is corundum. We also have "*Oriental amethyst*" and "*Oriental emerald*" for the purple, and the fine green, and "*Oriental aquamarine*" for the light blue-green corundum. The yellow corundum is also sometimes called "*King topaz,*" especially in Ceylon. Inferior sapphires of almost every conceivable color are frequently assorted in lots and sold as "fancy sapphires." Such lots, however, almost always need reclassification as they often contain as many as a dozen mineral species besides corundum.

Sapphires and rubies of minute tubular internal structure frequently display a beautiful six-pointed star when cut to a round-topped cabochon shape and exposed to direct sunlight or to light from any other single source. Such stones are named "*star sapphire*" and "*star ruby.*"

The artificial rubies and sapphires should all be called *scientific ruby* or *sapphire*, and not "*reconstructed*" or "*synthetic*" as none are made to-day from small, real rubies, and as the process is in no sense a chemical synthesis.

Chrysoberyl. Leaving the corundum gems we come next to chrysoberyl. When the gems furnished by this mineral are of a fine green by daylight, and of a raspberry red by artificial light, as is sometimes the case, they should be called "*Alexandrites*" (after the Csar Alexander II., in whose dominions, and on whose birthday, the first specimens are said to have been discovered). When chrysoberyl is of fibrous or tubular internal structure it affords cat's-eyes (when cabochon cut), and these should be specifically named as "*chrysoberyl cat's-eye*" to distinguish them from the less beautiful and less valuable quartz-cat's-eyes. Other varieties of chrysoberyl (most of those marketed are of a greenish yellow color) are correctly named simply "*chrysoberyls.*" Such stones are, however, sometimes incorrectly called "*chrysolite*" by the trade, and this practice should be corrected, as the term chrysolite applies correctly only to the mineral *olivine* which gives us the *peridot.*

Spinel. Next in the order that we have chosen comes "*spinel.*" The more valuable spinels are of a red color that somewhat closely approaches the red of some rubies. Such red spinels should be called "*Ruby spinel*" (and *not spinel ruby*). The stones themselves sometimes get mixed with corundum rubies (they are frequently found in the same gem gravels), and this makes it all the more necessary that both stones and names should be clearly distinguished. Some dealers call reddish spinels "*Balas ruby*" (rose red), and orange red ones "*rubicelle.*" Violet red spinel is sometimes called "*almandine spinel.*" It is very desirable that the name of the mineral species, *spinel*, should be used, together with a qualifying color adjective, in naming gems of this species, rather than such terms as "rubicelle," "balas-ruby," "spinel ruby," etc.

Topaz. We come now to *topaz.* True, or *precious topaz*, as it is usually called, to distinguish it from the softer and less valuable yellow quartz, is seldom seen in the trade to-day. Jewelers almost always mean yellow quartz when they speak of "topaz." This is an unfortunate confusion of terms, and one which will be hard to eradicate. There is seldom any injustice done through this misnaming, as the price charged is usually a fair one for the material offered. Considerably higher prices would be necessary if true topaz was in question.

An instance from the writer's experience will serve to illustrate the confusion that exists in the trade as to what should be called topaz. A jeweler of more than ordinary acquaintance with gems exhibited some fine brooch stones as specimens of topaz. On remarking that they were of course *citrine quartz* rather than *true topaz*, the author was met with the statement that the brooch stones were *real* topaz. In order to make clear to the dealer the difference between the two species, the author asked him if he hadn't some smaller topazes in stock that had cost him considerably more than the brooch stones. The dealer replied that he had some small wine yellow topazes for which he had paid more, and he produced them. The latter stones were true Brazilian topazes. Most of them had tiny, crackly flaws in them, as is frequently the case, and, as the writer pointed out to the dealer, they had been bought by the *carat*, whereas the large brooch stones had been bought at a certain price per *pennyweight.* In fact the little stones had cost more per carat than the larger ones had per pennyweight.

The dealer was then asked if there must not be some difference in the real nature of the two lots to justify paying more per carat for small, imperfect stones than per pennyweight for large perfect ones. He of course acknowledged that it would appear reasonable that such was the case. He was next shown that his small *true topazes* scratched his large stones easily, but the large ones could get no hold upon the surfaces of the small ones. (It will be remembered that topaz has a hardness of 8, while quartz has a hardness of 7.) The explanation then followed that the two lots were from two entirely distinct minerals, topaz and quartz, and that the former was harder, took a somewhat better polish, and was more rare (in fine colors) than quartz. Of course the yellow quartz should be sold under the proper name, *citrine quartz.* (From the same root that we have in "*citrus*" as applied to fruits. For example the "California Citrus Fruit Growers' Association," which sells oranges, lemons, grape fruit, etc. The color implication is obvious.) If the jeweler still wishes to use the term "topaz" because of the familiarity of the public with that name, then he should at least qualify it in some way. One name that is current for that purpose is "Spanish topaz," another is "Quartz-topaz." Perhaps the latter is the least objectionable of the names that include the word topaz.

Some of the wine yellow true topazes lose the yellow, but retain the pink component, on being gently heated. The resulting pink stone is rather pretty and usually commands a higher price than the yellow topazes. Such artificially altered topazes should be sold only for what they are, and probably the name "pinked topaz," implying, as it does, that something has been done to the stone, is as good a name as any. There is, however, little chance of fraud in this connection, as natural pink topazes are not seen in the trade, being very rare.

Some bluish green topaz is said to be sold as aquamarine, and this confusion of species and of names should, of course, be stopped by an actual determination of the material as to its properties. Lacking a refractometer, the widely differing specific gravities of the two minerals would easily serve to distinguish them.

LESSON XX

THE NAMING OF PRECIOUS STONES (*Concluded*)

Beryl, Emerald, Aquamarine. Coming now to *beryl* we have first *emerald*, then *aquamarine*, then beryls of other colors to consider. There is too often a tendency among dealers to confuse various green stones, and even doublets, under the name *emerald*. While the price charged usually bears a fair relation to the value of the material furnished, it would be better to offer tourmaline, or peridot (the mineral name of which is olivine), or demantoid garnet (sometimes wrongly called "Olivine"), or "emerald doublets," or emerald or "imitation emerald," as the case might be, under their own names.

There are no true "synthetic" or "seientific" or "reconstructed" emeralds, and none of these terms should be used by the trade.

There has been an effort made in some cases to do business upon the good reputation of the scientific rubies and sapphires, but the products offered, when not out and out glass imitations, have usually been doublets or triplets, consisting partly of some pale, inexpensive, natural mineral, such as quartz or beryl, and a layer of deep green glass to give the whole a proper color. All attempts to melt real emerald or beryl have yielded only a *beryl glass*, softer and lighter than true emerald, and not *crystalline*, but rather glassy in structure. Hence the names "reconstructed," "synthetic" and "scientific" should never be applied to emerald.

The light green and blue green beryls are correctly called *aquamarines*, the pale sky-blue beryls should be named simply *blue beryl*. Yellow beryl may be called *golden beryl*, or it may be called "*heliodor*," a name that was devised for the fine yellow beryl of Madagascar. Beautiful pink beryl from Madagascar has been called "*morganite*," a name that deserves to live in order to commemorate the great interest taken by J. Pierpont Morgan in collecting and conserving for future generations many of the gems in the American Museum of Natural History in New York.

Zircon. We now come to a number of minerals slightly less hard than beryl, but harder than quartz, and *zircon* is perhaps as hard as any of these, so it will be considered next. Red zircon, which is rare, is properly called "*hyacinth.*" Many Hessonite garnets (cinnamon stones) are incorrectly called hyacinths, however. The true hyacinth has more snap and fire owing to its adamantine surface luster and high dispersive power, as well as to its high refractive index. A true hyacinth is a beautiful stone. Golden yellow zircons are correctly called "*jacinths.*" Artificially whitened zircons (the color of which has been removed by heating) are known as "jargoons" or sometimes as "Matura diamonds." All other colors in zircon should be named simply zircon, with a color adjective to indicate the particular color as, "brown zircon," etc.

Tourmaline. Tourmaline furnishes gems of many different colors. These are all usually called simply tourmaline, with a color adjective to specify the particular color, as, for example, the "pink tourmaline" of California. Red tourmaline is, however, sometimes called "*rubelite*," and white tourmaline has been called "*achroite*." The latter material is seldom cut, and hence the name is seldom seen or used.

Garnet. We may next consider the *garnets*, as most of them are somewhat harder than quartz. As was said in Lesson XVIII. in our study of mineral species, there are several types of garnets, characterized by similarity of chemical composition, or at least by analogy of composition, but, having specific differences of property. The names used by jewelers for the several types of garnets ought to be a fairly true indication as to the type in hand in a particular case. At present there is considerable confusion in the naming of garnets. The most common practice is to call all garnets of a purplish red color "almandines." As many such garnets belong to the mineral species *almandite garnet*, there is little objection to the continuance of this practice. The somewhat less dense, and less hard blood red garnets are properly called "*pyrope garnets*" (literally "fire" garnets). Many of the Arizona garnets belong in this division. The term "Arizona *rubies*" should *not* be used. As was said under ruby, nothing but red corundum should receive that title. Similarly the pyrope garnet of the diamond mines of South Africa is incorrectly called "Cape ruby." Pyrope and almandite garnet tend to merge in composition and in properties, and the beautiful "*Rhodolite*" garnets of Macon County, North Carolina, are between the two varieties in composition, in color, and in other properties.

Hessonite garnet furnishes yellowish red and

brownish red stones, which are sometimes also called "cinnamon stones." They are also frequently and incorrectly called jacinth or hyacinth, terms which, as we have seen, should be reserved for yellow and red zircon, respectively.

Andradite garnet furnishes brilliant green stones, which have been incorrectly named "Olivines" by the trade. The name is unfortunate as it is identical with the true name of the mineral which gives us peridot. The name does not even suggest the color of these garnets correctly, as they are seldom olive green in shade. As the scarcity of fine specimens and their great beauty make a fairly high price necessary, the public would hardly pay it for anything that was called "garnet," as garnets are regarded as common and cheap. Perhaps the adoption of the name "*Demantoid*" might relieve the situation. The stones are frequently referred to as "demantoid garnets" on account of their diamond-like luster and dispersion. The use of "demantoid" alone, if a noun may be made from the adjective, would avoid both the confusion with the mineral olivine, and the cheapening effect of the word garnet, and would at the same time suggest some of the most striking properties of the material.

"*Spodumene*" furnishes pink to lilac "*Kunzite*," named after Dr. George F. Kunz, the gem expert, and for a time an emerald green variety was had from North Carolina which became known as "*Hiddenite*," after its discoverer, W. E. Hidden. No confusion of naming seems to have arisen in regard to this mineral.

The next mineral in the scale of hardness is quartz. (Hardness 7.) When pure and colorless it should be called "*rock crystal.*" Purple quartz is of course *amethyst*. Some dealers have adopted a bad practice of calling the fine deep purple amethyst "Oriental" amethyst, which should not be done, as the term "Oriental" has for a long time signified a *corun-*

dum gem. As Siberia has produced some very fine amethysts, the term "*Siberian amethyst*" would be a good one to designate any especially fine gem.

Quartz Gems. We have already considered the naming of yellow quartz in connection with topaz. "*Citrine quartz*" is probably the best name for this material. If it is felt that the name "topaz" must be used, the prefix "quartz" should be used, or perhaps "Spanish topaz" will do, but some effort should be made to distinguish it from the true precious topaz. In addition to amethyst and citrine quartz we have the pinkish, milky quartz known as "*rose quartz.*" This is usually correctly named.

"*Cat's-eye*" is a term that should be reserved for the Chrysoberyl variety, and the quartz variety should always be called "quartz-cat's-eye." "*Tiger's-eye*" is a mineral in which a soft fibrous material has been dissolved away, and quartz has been deposited in its place. "*Aventurine quartz*" is the correct name for quartz containing spangles of mica. Clear, colorless pebbles of quartz are sometimes cut for tourists. Such pebbles are frequently misnamed "diamonds" with some prefix, as for example "Lake George diamonds," etc. Among the minutely crystalline varieties of quartz we have the clear red, which should be called "*carnelian*," the brownish red "*sard*," the green "*chrysoprase*," the leek green "*prase*," and the brighter green "*plasma*." The last three are not so commonly seen as the first two, and frequently the best-colored specimens are artificially dyed.

"*Jasper*," a material more highly regarded by the ancients than at present, is mainly quartz, but contains enough earthy material to make it opaque. "*Bloodstone*" is a greenish chalcedony with spots of red jasper.

"*Agates*" are banded chalcedonies, the variety called "*onyx*" having very regular bands, and the "*sardonyx*" being an onyx agate in which some of the bands are of reddish sard.

Just as we considered opal with quartz (because of its chemical similarity) when discussing mineral species, so we may now consider the proper naming of opals here. "*Precious opal*" is distinguished from "*common opal*" by the beauty of its display rather than by any difference in composition. The effect is of course due to the existence of thin films (probably of material of slightly different density), filling what once were cracks in the mass. The rainbow colors are the result of interference of light (see a college text on physics for an explanation of interference). The varying thickness of these films gives varying colors, so different specimens of opal show very different effects. The differences of distribution of the films within the material also cause variations in the effects. Hence we have hardly any two specimens of opal that are alike.

There are, however, certain fairly definite types of opal and jewelers should learn to apply correct names to these types. Most prominent among the opals of to-day are the so-called "*Black opals*" from New South Wales. These give vivid flashes of color out of seeming darkness. In some positions the stones, as the name implies, appear blue-black or blackish-gray. By transmitted light, however, the bluish stones appear yellow. Owing to the sharp contrast between the dark background and the flashing spectrum colors, black opals are most attractive stones and fine specimens command high prices. One fine piece, which was on exhibition at the Panama-Pacific Exposition was in the shape of an elongated shield, about 1¾ inches by 1⅛ inches in size and rather flat and thin for its spread. It gave in one position a solid surface of almost pure ruby red which changed to green on tipping the stone to the opposite direction; $2,000 was asked for the piece.

"*White opal*" is the name applied to the lighter shades of opal which do not show the bluish-black effect in any position. "*Harlequin opal*" has rather large areas of definite colors giving somewhat the effect of a map of the United States in which the different States are in different colors.

"*Fire opal*" is an orange-red variety. It has some "play" of colors in addition to its orange-red body color.

"*Opal Matrix*" has tiny specks and films of precious opal distributed through a dark volcanic rock and the mass is shaped and polished as a whole.

Jade. "*Jade*" should next receive attention. It is a much abused term. Under it one may purchase *jadeite*, *nephrite*, *bowenite*, *amazonite*, or frequently simply *green glass*. The use of the word ought to be confined to the first two minerals mentioned, namely, jadeite and nephrite, for they only possess the extreme toughness together with considerable hardness that we expect of jade. Bowenite, while tough, is relatively soft and amazonite is brittle and also easily cleavable, while glass is both soft and brittle.

Peridot and Olivine. The mineral "*olivine*" gives us the "*peridot*" (this name should be kept for the deeper bottle green stones), and the olive green gems of this same mineral may correctly be called "*olivine*" or "*chrysolite.*" As was explained under garnet, jewelers frequently use the term "olivine" to designate demantoid garnet. The term chrysolite is also sometimes incorrectly used for the greenish-yellow chrysoberyl.

Feldspar Gems. Among the minerals softer than quartz, which are used as gems, we have also "*feldspar,*" which gives us "*moonstone,*" "*Labradorite,*" and "*Amazonite.*"

An opalescent form of chalcedony is frequently gathered on California beaches and polished for tourists under the name of "*California Moonstone.*" This name is unfortunately chosen as the material is not the same as that of true moonstone and the effect is not so pronounced or so beautiful. The polished stones show merely a milky cloudiness without any of that beautiful sheen of the true moonstone. "*Labradorite*" is usually correctly named. "*Amazonite*" was originally misnamed, as none is found along the river of that name. The term has come into such general use, however, that we shall probably have to continue to use it, especially as no other name has come into use for this bluish-green feldspar. As has already been said, amazonite is sometimes sold as "jade," which is incorrect.

Malachite, Azurite, and Lapis Lazuli. *Malachite* and *azurite* are usually correctly named, but "*lapis lazuli*" is a name that is frequently misused, being applied to crackled quartz that has been stained with Prussian blue, or some other dye, to an unconvincing resemblance to true lapis. Such artificially produced stones are sometimes sold as "*Swiss lapis.*" They are harder than true lapis and probably wear much better in exposed ornaments, but they are not lapis and are never of equal color, and names should not be misused, and especially is this true in a trade where the public has had to rely so completely upon the knowledge and the integrity of the dealer.

With the increase of knowledge about precious stones that is slowly but steadily growing among the public, it becomes more than ever necessary for the jeweler and gem dealer to know and to use the correct names for all precious stones. The student who wishes to learn more about the matter will have to cull his information from many different works on gems. G. F. Herbert-Smith, in his *Gem Stones*, gives a three and one half page chapter on "Nomenclature of Precious Stones" (Chap. XIII, pp. 109–112). The present lesson has attempted to bring together in one place material from many sources, together with some suggestions from the author.

LESSON XXI

WHERE PRECIOUS STONES ARE FOUND

Occurrence of Diamond. Every dealer in precious stones should know something of the sources of the gems that he sells. The manner of the occurrence of the rough material is also a matter of interest. It will therefore be the purpose of this lesson to give a brief account of the geographical sources of the principal gems and of their mode of occurrence in the earth.

For the sake of uniformity of treatment we will once more follow the descending order of hardness among the gems and we thus begin by describing the occurrence of diamond. It will be of interest to note first that the earliest source of the diamond was India, and that for many years India was almost the sole source. Tavernier tells us that the diamond mining industry was in a thriving state during the years from 1640 to 1680, during which time he made six journeys to India to purchase gems. He speaks of Borneo as another source of diamonds, but most of the diamonds of that time were furnished by India.

"Golcondas." Indian diamonds were noteworthy for their magnificent steely blue-white quality and their great hardness, and occasionally one comes on the market to-day with an authentic pedigree, tracing its origin back to the old Indian mines, and such stones usually command very high prices. One of a little over seven and one half carats in weight, in the form of a perfect drop brilliant, has lately been offered for sale at a price not far from $1,000 per carat. Such diamonds are sometimes called "Golcondas" because one of the mining districts from which the fine large Indian stones came was near the place of that name. Some of the stones from the Jägersfontein mine in South Africa resemble the Golcondas in quality. Many of the large historical crown diamonds of Europe came from the Indian mines.

The stones were found in a sedimentary material, a sort of conglomerate, in which they, together with many other crystalline materials, had become imprisoned. Their original source has never been determined. They are therefore of the so-called "River" type of stone, having probably been transported from their original matrix, after the disintegration of the latter, to new places of deposit, by the carrying power of river waters.

The Indian mines now yield very few stones. The United States Consular reports occasionally mention the finding of a few scattered crystals but the rich deposits were apparently worked out during the seventeenth century and the early part of the eighteenth century.

In 1725 and in the few following years the Brazilian diamond fields began to supersede those of India. Like the latter, the Brazilian fields were alluvial, that is, the materials were deposited by river action after having been carried to some distance from their original sources.

Brazilian Diamonds. The diamonds of Brazil also resembled those of India in quality, being on the average better than those of the present South African mines. It may be added that even the African diamonds that are found in "river diggings" average better in quality than those of the volcanic pipes which form the principal source of the world's supply to-day, There seems to be a superabundance of iron oxide in the rocks of the African mines and in the diamonds themselves, imparting yellow or brownish tints to the material. The "River" stones seem to have lost this color to a considerable extent, if they ever had it. Possibly long extraction with water has removed the very slightly soluble coloring material. Whatever the cause of their superiority "River" stones have always been more highly regarded than stones from the volcanic pipes.

Brazil furnished the world's principal supply of diamonds until the discovery of the African stones in 1867. At present relatively small numbers of Brazilian stones reach the world's markets. Most of these come from the great Bahia district (discovered in 1844) rather than from the older mines of Brazil. The present Brazilian stones average of small size. They are, however, of very good quality as a rule. A few green stones are found in Brazil and these may be of an absinthe-green or of a pistachio-green tint.

Australian and American Sources. While a few diamonds now come on the market from New South Wales, and while an occasional stone is found in the United States (usually in glacial drift in the north central States, or in volcanic material somewhat resembling that of South Africa in Arkansas) yet the world's output now comes almost entirely from South Africa and mainly from the enormous volcanic pipes of the Kimberly district and those of the Premier Co. in the Transvaal.

South African Diamonds. The nature of the occurrence of diamond in the "pipes" of South Africa is so well known to all who deal in diamonds to-day that but little space need be devoted to it. The "blue ground," as the rock in which the diamonds are found is called, seems to have been forced up from below, perhaps as the material of a mud volcano, bringing with it the diamonds, garnets, zircons, and the fifty or more other minerals that have been found in the blue ground. The fragmentary character of some of these minerals would indicate that the blue ground was not their original matrix. How the diamonds originally crystallized and where, is still probably a matter for further speculation.

While at first the mines were worked, like quarries, from the surface, and while the great Premier mine is still so worked, most of the present mines are worked by sinking shafts in the native rock outside of the blue ground and then tunneling into the diamond-bearing rock laterally, removing it to the surface, allowing it to weather on the "floors" until it crumbles, then crushing and washing it and concentrating the heavy minerals by gravity methods. Large diamonds are then picked out of the concentrates by hand and small ones and fragments are removed by the "greasers," which are shaking tables heavily smeared with grease over which the concentrates are washed and to which diamond alone, of all the minerals in the concentrate, sticks. The grease is periodically removed and melted, and the diamonds secured. The grease can then be used again.

German South West Africa furnishes a considerable output of very small diamonds, which are found in dry sand far from any present rivers. These diamonds cut to splendid white melée and the output is large enough to make some difference in the relative price of small stones as compared to large ones. The South West African field seldom yields a stone that will afford a finished quarter-carat diamond.

Rubies. Passing on to the occurrence of the *corundum* gems we will consider first the *ruby*. Most fine rubies come from Burmah. The district in which they are found is near Mogok. Practically all the fine pigeon-blood rubies come from this district. The fashion for red stones being for the time little in evidence rubies are not now in great demand. This cessation of demand can hardly be laid to the competition of the scientific ruby, for the sapphire is now very much in vogue, yet scientific sapphires resemble the natural ones even more closely than do the rubies.

Siam furnishes a considerable number of dark garnet-like rubies. These do not command high prices. They are, however, sometimes very beautiful, especially when well cut for brilliancy, and when in a strong light.

Ceylon furnishes a few rubies and a few red corundums have been found in North Carolina.

The Burmese rubies appear to have been formed in a limestone matrix, but most of those obtained are gotten from the stream beds, where they have been carried by water after weathering out from the mother rock.

The rubies of Ceylon, too, probably originated in a limestone matrix, but are sought in stream gravels.

Sapphires. Fine blue sapphires originate in Siam in larger numbers than in any other locality. Kashmir, in India, also supplies splendid specimens of large size. Ceylon, too, furnishes a good deal of sapphire, but mostly of a lighter color than the Kashmir sapphire. The Ceylon sapphires are found in the streams, but originate in rock of igneous origin.

Montana furnishes considerable quantities of sapphire, some of which is of very good color. It is, of course, as good as the Oriental if of equal color, being of the same material. The better colored sapphire from Montana is mined from the rock. Most of the sapphires found in the river gravels near Helena, Mont., are greenish-blue or of other colors, and not of fine blue.

Queensland and Victoria in Australia supply considerable quantities of sapphire. When blue the Australian sapphire is usually too dark to be very valuable. The golden and other "fancy" sapphires of the trade come largely from the Ceylon gravels. Siam yields silky brown stones and some fine green ones. Some of the Australian sapphires when cut in certain directions yield green stones.

Chrysoberyl. Chrysoberyl of the variety Alexandrite now comes mainly from Ceylon, although formerly from the Ural Mountains.

The cat's-eyes also come chiefly from Ceylon.

The yellowish-green chrysoberyls (which jewelers sometimes call chrysolite) come both from Ceylon and from Brazil. They are frequently found in papers of "fancy sapphires" or "fancy color stones," so called.

Spinel. Spinels are found along with ruby in Burmah and in Siam and they also occur in the gem gravels of Ceylon. Limestone is the usual matrix of spinel, although it is more often mined in gravels resulting from the weathering of the matrix.

Topaz. True topaz, of wine-yellow color, comes mostly from Brazil. Ceylon also furnishes yellow topaz. Asiatic Russia furnishes fine large blue or blue-green crystals resembling aquamarine in appearance. Most of the topaz found in other localities is pale or colorless. Several of our western States, notably Utah, Colorado, and California, furnish colorless topaz. Mexico and Japan also produce it. It is seldom cut, for, while producing a rather brilliant stone, it has little "fire" and is therefore not very attractive.

Emerald and Aquamarine. Beryl of the emerald variety is exceedingly scarce in the earth. Most of the best emerald comes from Colombia, South America. Large crystals of paler color come from the Urals.

Like ruby and spinel, emerald usually originates in limestone. One is tempted to suspect that these stones are of aqueous origin and that sapphires, and beryl, other than emerald, are more likely of igneous origin.

Beryls of the aquamarine type occur in many places, but usually of too pale a tint, or too imperfect, to be worthy of cutting. Fine gem beryl of blue and blue-green tints comes from Siberia and from several places in the Ural Mountains on their Asiatic slopes.

The Minas Garaes district of Brazil, famous for all kinds of gem stones, furnishes most of the aquamarine of commerce. The pegmatite dikes of Haddam Neck, Conn., of Stoneham, Me., and of San Diego County, Cal., have furnished splendid aquamarine and other beryl. These dikes, according to the geological evidence, are the result of the combined action of heat and water. Thus both melting and dissolving went on together and as a result many fine gem minerals of magnificent crystallization were formed during the subsequent cooling. The longer the cooling lasted and the more free space for growth the crystals had, the larger and more perfect they got. The author has himself obtained finely crystallized aquamarine and tourmaline from the Haddam, Conn., locality and the best specimens there occur in "pockets" or cavities in the coarse granite. Within, these pockets are lined with crystals of smoky quartz, tourmaline, beryl, and other minerals. Sometimes crystals occur in mud or clay masses inside the cavities and such crystals, having been free to grow uninterruptedly in every direction, were perfect in form, being doubly terminated, and not attached anywhere to the rock.

Madagascar has in recent years furnished the finest pink beryl, which has been named Morganite. Yellow beryl (Heliodor) and aquamarine also occur in Madagascar.

Zircon. Zircon comes on the market mainly from Ceylon. It deserves to be as much esteemed in this country as it is in Ceylon, for its optical properties are such that it is a very snappy stone. Some of the colors in which it occurs, such as the golden browns, lend themselves nicely to the matching of gems and garments, and, with the growth of education in such matters, jewelers would do well to get better acquainted with the possibilities of zircon and to introduce it to their customers. The supply from Ceylon is sufficient to justify popularizing the stone. Small zircons are found in almost every heavy concentrate, as, for example, in the concentrates of the diamond mines of South Africa, and in those of gold placers in many places. The rough stones resemble rough diamonds in luster and are sometimes mistaken for diamonds.

Garnets. Garnets of various types are found widely distributed in nature. Perhaps the Bohemian supply is best known, having furnished a host of small stones which have usually been rose cut for cluster work or made into beads. The Bohemian garnets are of the pyrope or fire-red type. Relatively few large stones of sufficient transparency for cutting are produced in the Bohemian mines. The so-called "Cape rubies" of the diamond mines of South Africa are pyrope garnets and some large and fine ones are found. The "Arizona rubies" are pyrope garnets, and while seldom of notable size, some are of very fine color, approaching deep rubies, and the color remains attractive by artificial light.

Almandite garnet, the "almandine" of the jeweler is less abundant than pyrope, when of gem quality. Ceylon furnishes some and India furnishes perhaps more. Brazil, from its prolific gem gravels at Minas Novas, supplies good almandite, and smaller quantities are found in many different localities.

Hessonite garnet, the cinnamon stone or "hyacinth" (incorrect) of the trade, comes mainly from Ceylon.

Andradite garnet, of the variety known as demantoid, from its diamond-like properties, and which is usually sold under the misleading name "olivine" in the trade, comes from the western slopes of the Ural Mountains.

Tourmaline. Gem tourmaline comes from Ceylon, from Madagascar, from the Ural Mountains, from Brazil, from Maine, from Connecticut, and from California.

The Ceylon tourmalines are mostly yellow or yellowish-green, sometimes fine olive-green. Those from the Urals may be pink, blue or green. Brazilian tourmalines are usually green, but sometimes red. In fact in many localities several colors of tourmaline are usually found together and it may be that a single crystal will be green in most of its length but red or pink tipped. Some crystals have a pink core and a green exterior. The author has found both of the two latter types in the Haddam, Conn., tourmalines, and on one occasion was surprised to get back a wine-colored tourmaline from a cutter to whom he had sent a green crystal. There was but a thin shell of the green material on the outside of the crystal.

Some of the Madagascar tourmaline is of a fine brownish-red, almost as deep as a light garnet, and much clearer than most garnet.

Would it not be fitting on account of its occurrence in several localities in the United States, for Americans to use more tourmaline in their jewels? The quality of some of the tourmalines of Maine, and of California especially, is not excelled by tourmaline from any other locality. Some of the Maine tourmaline is of a delightful, slightly bluish-green tint that almost approaches emerald.

Kunzite. Spodumene, of the variety kunzite, comes from San Diego County, California.

Quartz Gems. Coming now to the quartz gems we find amethyst and citrine, or golden quartz widely distributed so that only the localities that furnish the better grades of these stones need be mentioned. Siberia and Uruguay furnish fine amethyst. Brazil also furnishes large quantities of very good quality.

Amethyst. The chief charm of the Siberian amethyst lies in its large red component, which enables it to change from a deep grape-purple by daylight to a fine red by artificial light that is rich in red rays, and poor in blue ones. The paler types of amethysts that were once esteemed, probably for lack of the rich deep variety, become gray in appearance and much less lovely under artificial light. India furnishes some amethysts, and papers of "fancy color stones" containing native cut gems from Ceylon, frequently contain amethysts, but Brazil, Uruguay, and Siberia furnish the great bulk of the stones that are regarded as choice to-day.

Yellow Quartz. Citrine or golden quartz comes mainly from Brazil. The "Spanish topaz" is sometimes the result of heating smoky quartz from Cordova province in Spain. Our own western mountains furnish considerable yellow and smoky quartz fit for cutting.

Rose Quartz. Rose quartz of the finest quality comes from South Dakota. Bavaria, the Ural Mountains, and Paris, Maine, have also furnished it.

Agate. Agates of the finest types, such as carnelian and sard, come principally from Brazil and from India.

Opal. Opals now come most largely from Australia, the Hungarian mines yielding but few stones at present. The fine black opals of New South Wales are unsurpassed by any that have ever been found elsewhere, Mexico furnishes considerable opal, and is notable for its fine "fire opal" or "cherry opal."

Jade. Most of the jade of the variety nephrite that is obtained to-day comes from several of the provinces of China or from Siberia or from Turkestan. A dark-green nephrite comes from New Zealand.

Jade of the jadeite variety, which is harder than nephrite and more highly valued, is rare. The best specimens come from Upper

Burmah. It is also found in China and in Tibet.

Peridot. Peridot, and the brighter olivine or chrysolite, while of the same mineral species, do not seem to occur together. The darker bottle-green specimens come from the Island of St. John in the Red Sea. It is said that many of the finer peridots now available have been recut from old stones mined many years ago. Queensland supplies light-green chrysolite, and Arizona a yellowish-green variety. Light-green stones have been found near the ruby mines of Upper Burmah.

Moonstone. Moonstone comes mainly from Ceylon. The native cut specimens are sent here and recut, as, when native cut, the direction of the grain is seldom correct to produce the moonlight effect in symmetrical fashion. The native cutters apparently try to retain all the size and weight that is possible, regardless of the effect.

Turquoise. Turquoise of the finest blue and

most compact texture (and hence least subject to color change) comes from the province of Khorassan in Persia. Several of our western states supply turquoise of fair quality, notably New Mexico, Arizona, Nevada, and California.

Lapis Lazuli. Lapis Lazuli comes from Afghanistan, from Siberia, and from South America.

Malachite. Malachite is found in many copper mines, but principally in those of the Ural Mountains.

Azurite. Azurite is found in the Arizona mines and in Chessy, in France (hence the name chessylite, sometimes used instead of azurite).

References. Students who wish to get a fuller account of the occurrence of precious stones should run through G. F. Herbert-Smith's *Gem-Stones* under the different varieties. This work is the most recent authentic work of a strictly scientific character. Dr. George F. Kunz's *Gems and Precious Stones of North*

America gives a detailed account of all the finds in North America up to the time of publication. Many of these are of course of little commercial importance. The *Mineral Resources of the United States* contains annually a long account of the occurrences of gem materials in this country. A separate pamphlet containing only the gem portion can be had gratis from the office of the United States Geological Survey, Washington, D. C.

LESSON XXII

HOW ROUGH PRECIOUS STONES ARE CUT

Rough Precious Stones. John Ruskin, who had the means to acquire some very fine natural specimens of gem material was of the opinion that man ought not to tamper with the wonderful crystals of nature, but that rather they should be admired in the rough. While one can understand Ruskin's viewpoint, nevertheless the art of man can make use of the optical properties of transparent minerals, properties no less wonderful than those exhibited in crystallization, and indeed intimately associated with the latter, and, by shaping the rough material in accordance with these optical properties, greatly enhance the beauty of the gem.

No material illustrates the wonderful improvement that may be brought about by cutting and polishing better than diamond. In the rough the diamond is less attractive in appearance than rock crystal. G. F. Herbert-Smith likens its appearance to that of soda crystals. Another author likens it to gum arabic. The surface of the rough diamond is usually ridged by the overlapping of minute layers or strata of the material so that one cannot look into the clear interior any more than one can look into a bank, through the prism-glass windows that are so much used to diffuse the light that enters by means of them. Being thus of a rough exterior the uncut diamond shows none of the snap and fire which are developed by proper cutting.

As the diamond perhaps shows more improvement on being cut than any other stone, and as the art of cutting the diamond is distinct from that of cutting other precious stones, both in the method of cutting and in the fact that the workers who cut diamonds cut no other precious stones, it will be well to consider diamond cutting separately.

Before discussing the methods by which the shaping and polishing are accomplished let us consider briefly the object that is in view in thus altering the shape and smoothing the surface of the rough material.

How Cutting Increases Brilliancy. Primarily the object of cutting a diamond is to make it more brilliant. So true is this that the usual form to which diamonds are cut has come to be called the *brilliant.* The adjective has become a noun. The increased brilliancy is due mainly to two effects: First, greatly increased reflection of light, and second, dispersion of light. The reflection is partly external but principally internal.

Taking up first the internal reflection which is responsible for most of the white brilliancy of the cut stone we must note that it is a fact that light that is passing through any transparent material will, upon arriving at any polished surface, either penetrate and emerge or else it will be reflected within the material, depending upon the angle at which the light strikes the surface. For each material there is a definite angle outside of which light that is passing as above

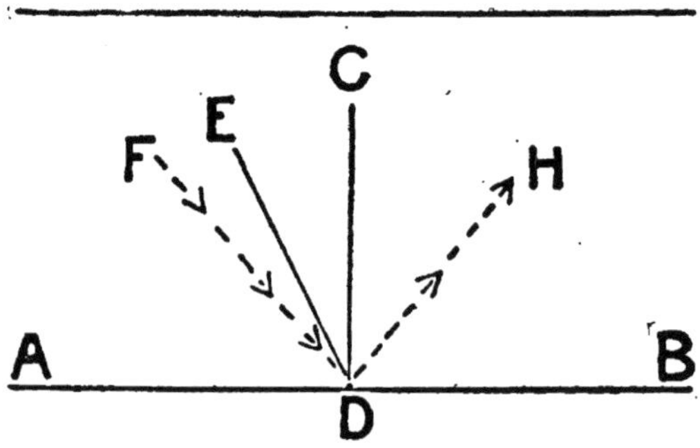

FIG. 9

A B represents the back surface of a piece of diamond.
C D is a line perpendicular to *A B*.
Angle *C D E* is about 24 degrees.
Dotted line, *F D H* represents the course taken by a ray of light which is totally reflected at *D* in such fashion that angle *F D A* equals angle *H D B*.
Any light proceeding towards *A B* but between *E* and *C*, would fail to be totally reflected. Most of it would penetrate *A B*.

described, is *totally reflected* within the material.

Total Reflection. For diamond this *critical angle*, as it is called, is very nearly 24° from a perpendicular to the surface. If now, we shape a diamond so that most of the light that enters it from the front falls upon the first back surface that it meets, at an angle greater than 24°

to a perpendicular to that surface, the light will be totally reflected within the stone. The angle at which it is reflected will be the same as that at which it meets the surface. In other words the angles of incidence and of reflection are equal. See Fig. 9 for an illustration of this point.

Theory of the "Brilliant." In the usual "brilliant" much of the light that enters through the front surface is thus totally reflected from the first rear facet that it meets and then proceeds across the stone to be again totally reflected from the opposite side of the brilliant. This time the light proceeds toward the top of the stone. See Fig. 10—(From G. F. Herbert-Smith's *Gem Stones*).

The angles of the top of a brilliant are purposely made so flat that the up coming light fails to be totally reflected again and is allowed to emerge to dazzle the beholder. In the better made brilliants the angle that the back slope makes with the plane of the girdle is very nearly 41° and the top angle, or angle of the front slope to the plane of the girdle is about 35°. Such well made brilliants when held up to a bright light appear almost black—that is,

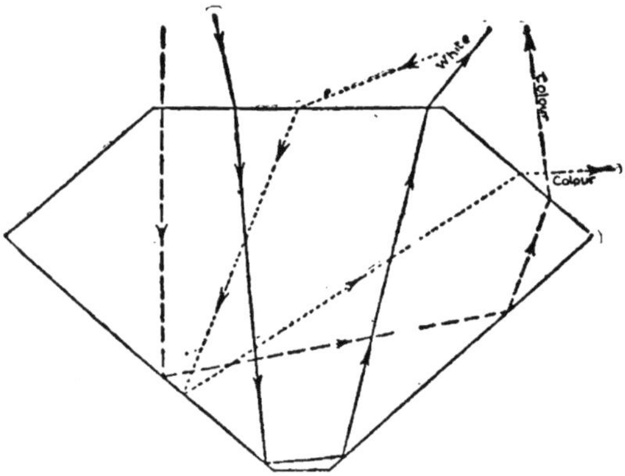

FIG. 10.—COURSE OF THE RAYS OF LIGHT PASSING THROUGH A BRILLIANT.

they fail to pass any of the light through them (except through the tiny culet, which, being parallel to the table above, passes light that comes straight down to it).

In other words, instead of allowing the light to penetrate them, well-made brilliants almost totally reflect it back toward its source, that is, toward the front of the stone. The well-cut diamond is a very brilliant object, viewed from the front.

We must now consider how the "fire" or prismatic color play is produced, for it is even more upon the display of fire than upon its pure white brilliancy that the beauty of a diamond depends.

Cause of "Fire." As we saw in Lesson X (which it would be well to reread at this time), white light that changes its course from one transparent medium to another at any but a right angle to the surface involved, is not only refracted (as we saw in Lesson II) but is dispersed, that is, light of different colors is bent by differing amounts and thus we have a separation of the various colors. If this takes place as the ray of light leaves the upper surface of a brilliant the observer upon whose eye the light falls will see either the red, or the yellow, or the blue, as the case may be, rather than the white light which entered the stone. If instead, the dispersion takes place as the light enters the brilliant the various colored rays thus produced will be totally reflected back to the observer (slightly weakened by spreading, as compared to the direct or unreflected spectra). Thus dispersion produces the "fire" in a brilliant.

Other materials than diamond behave similarly, but usually to a much smaller extent, for few gem materials have so high a refractive power or so great a dispersive power as diamond.

Having considered the theory of the brilliant we may now take up a study of the methods by which the exceedingly hard rough diamond is shaped and polished.

Cleaving Diamonds. If the rough material is of poor shape, or if it has conspicuous defects in it which prevent its being made into a single stone, it is cleaved (*i. e.*, split along its grain). Hard as it is, diamond splits readily in certain definite directions (parallel to any of the triangular faces of the octahedral crystal). The cleaver has to know the grain of rough diamonds from the external appearance, even when the crystals, as found, are complicated modifications of the simple crystal form. He can thus take advantage of the cleavage to speedily reduce the rough material in size and shape to suit the necessity of the case. The cleaving is accomplished by making a nick or groove in the surface of the rough material at the proper point (the stone being held by a tenacious wax, in the end of a holder, placed upright in a firm support). A thin steel knife blade is then inserted in the nick and a sharp light blow struck upon the back of the knife blade. The diamond then readily splits.

"Cutting Diamonds." The next step is to give the rough material a shape closely similar to that of the finished brilliant but rough and without facets. This shaping or "cutting" as it is technically called, is done by placing the rough stone in the end of a holder by means of a tough cement and then rotating holder and stone in a lathe-like machine. Another rough diamond (sometimes a piece of bort, unfit for cutting, and sometimes a piece of material of good quality which it is necessary to reduce in size or alter in shape) is cemented into another holder and held against the surface of the rotating diamond. The holder is steadied against a firm support. It now becomes a case of "diamond cut diamond," each stone wearing away the other and being worn away itself.

The cutting process is fairly rapid and it leaves the stone (which is reversed to make the opposite side) round in form and with a rounding top and cone shaped back. Stones of fancy shape, such as square, or cushion shape, have to be formed in part by hand rubbing or "bruting" as it is called.

The facets must now be polished onto the stone. Usually the workers who cut do not cleave or polish.

"Polishing" Diamonds. The polisher fixes the cut stone firmly in a metallic holder called a dop, which is cleverly designed to hold the stone with much of one side of it exposed. The holder is then inverted so that the stone is beneath and a stout copper wire attached to the holder is then clamped firmly in a sort of movable vise. The latter is then placed on the bench in such a position that the diamond rests upon the surface of a rapidly revolving horizontal iron wheel or "lap" as it is called. The surface of the latter is "charged" with diamond dust, that is, diamond dust has been pushed into the metal surface which thus acts as a support to the dust. The latter wears away the diamond, producing a flat facet. The lap is kept moistened with oil and from time to time fresh oil and diamond dust are applied. A speed of about 2,000 rotations per minute is used.

Facetting. The making of the facets is rather slow work, especially when, as is usually the case in making the "table" the work has to be done against one of the "hard points" of the crystal. Great care has to be taken to place the stone so that the grain lies in a correct position, for diamond cannot be polished against the grain, nor even exactly with it, but only obliquely across it. This requirement, as much as anything, has prevented the use of machines in polishing diamonds. The table is usually first polished on, then the four top slopes, dividing the top surface into quarters, then each of the four ridges thus left, is flattened, making eight facets and finally 32 facets, exclusive of the table, are made upon the top of the brilliant. The stone is then reversed and 24 facets, and the culet, polished on the back. As each facet nears its proper shape the stone is placed upon a particularly smooth part of the lap and a slight vibratory motion given to the holder by the hand. This smooths out any lines or grooves that may have formed because of inequalities of surface of the lap. When completely facetted the brilliant is finished and requires only to be cleaned, when it is ready for sale.

LESSON XXIII

HOW ROUGH PRECIOUS STONES ARE CUT AND WHAT CONSTITUTES GOOD "MAKE"—*Concluded*

Slitting and Cleaving. The cutting and polishing of precious stones other than diamond is a trade entirely distinct from diamond cutting. The precious stone lapidary cuts every species of stone except diamond. The methods used by different lapidaries vary somewhat in their details, and there are many trade secrets which are more or less jealously guarded by their possessors, but in general the methods used to reduce the rough materials to the finished gems are as follows: First, the rough material, if of too large size, or if very imperfect, is *slitted*, or, if it possesses a pronounced cleavage, it may be *cleaved*, in order to reduce the size or to remove imperfect parts. *Slitting* is accomplished by means of a circular disc of thin metal which is hammered so that it will be flat and rotate truly, and is then clamped between face plates, much as an emery wheel is held. The smooth edge of the circular disc is then charged with diamond dust and oil, the diamond dust being bedded into the edge of the metal disc by the pressure of some hard, fine-grained material, such as chalcedony, or rolled into the metal by the use of a rotating roller. Once charged, and kept freely supplied with oil, a slitting wheel will slice a considerable number of pieces of any precious stone less hard than diamond, and will do so with considerable rapidity. The wheel is, of course, rotated very rapidly for this purpose.

The cleaving of certain gem materials, such as true topaz (which splits perfectly across the prism, parallel to its base) is easily accomplished, and it is done in much the same manner as the cleaving of diamond. The feldspar gems, such as moonstone, amazonite, and labradorite, also cleave very smoothly in certain directions. Spodumene, of which Kunzite is a variety, cleaves almost too easily to be durable. Most gem minerals, however, lack such perfect cleavage and when it is desired to remove imperfect parts, or to reduce large pieces to smaller sizes, these materials are slitted as above described.

"Rubbing Down." The material being of nearly the dimensions of the finished piece, the next step is to "rub it down," as it is called, to approximately the shape and size desired. This rubbing down process was formerly done by means of a soft metal lap (sometimes of lead), charged with coarse emery powder and water. Carborundum, being harder and sharper than emery, has replaced it very largely. Some of the softer materials, such, for example, as turquoise, are rubbed down on a fast flying carborundum wheel of similar type to those used in machine shops for grinding steel tools. These wheels rotate in a vertical plane and are kept wet. The laps before mentioned run horizontally. The carborundum wheels have the grains of carborundum cemented together by means of some binding material and this gradually crumbles, exposing fresh, sharp cutting edges. Various sizes of grain, and various degrees of hardness of the binding material, as well as various speeds, are needed to suit the many different materials rubbed down by the lapidary. Some lapidaries rub down the harder and more valuable gems such as ruby upon diamond charged laps of brass or other metal.

Cabochons. The rubbing down process does not leave a facetted surface, but only a coarse roughly rounded or flattened surface. If the material is to be left in some one of the flat-backed, rounded top forms known as cabochon cut, the surfaces need only to be smoothed (by means of very fine abrasives such as fine emery applied by means of laps, or even by fine emery or carborundum cloth), and they are then ready for polishing.

Facetted Stones. If, however, the stone is to be facetted in either the brilliant form, somewhat like the diamond, or step cut or otherwise facetted, it is cemented strongly onto a holder (much like the wooden part of a pen holder). The upper end of the holder is rested in one of a series of holes in what is called a "*ginpeg*" resting in the work-bench near a metal lap, and the stone is pressed upon the rapidly rotating surface of the lap, which is charged with diamond dust or carborundum, according to the hardness of the material to be facetted. A flat facet is thus ground upon the stone. By rotating the holder a series of facets, all in the same set, is produced. The holder is then changed to a new position on the ginpeg and another set of facets laid upon the stone. Thus as many as four or five tiers or sets of facets may be applied to one side, say the top of the stone. The latter is then removed, from the holder and cemented to it again, this time with the bottom exposed, and several sets of facets applied.

The stone is now *cut* but not *polished*. The facets are flat, but have a rough ground-glass like surface. The polishing is usually done by workers who do not cut stones, but who do nothing but polish them. In small shops, however, the same lapidary performs all the parts of the work.

Polishing. The polishing of stones, whether cabochon or facetted, is accomplished by the use of very finely powdered abrasives such as corundum powder, tripoli, pumice, putty powder, etc. Each gem material requires special treatment to obtain the best results. It is here that most of the trade secrets apply.

The troubles of the lapidary in getting the keen polish that is so much admired on fine gems are many. In general, the polishing powder should not be quite as hard as the material to be polished, else it may grind rather than polish. The material should be used with water or oil to give it a creamy consistency. It should be backed by laps of different materials for different purposes. Thus, when backed by a fairly hard metal even tripoli, although much softer, will polish sapphire. On a lap of wood, tripoli would fail to polish hard materials, but would polish amethyst or other quartz gem. A change of speed of the lap, too, changes the effect of the polishing material. I have seen a lapidary, who was having no success at polishing an emerald, get very good results by using a stick as a brake and slowing down his lap.

The polishing material must be of very uniform size, preferably water floated or oil floated, to give good results. The lap must be kept flat and true and the stone must be properly held, or the flatness of the facets, upon which brilliancy depends in part, will be destroyed during the polishing.

The softer materials, such as opal, require treatment more like that accorded cut glass, and soft abrasive powders, such as pumice, suffice to polish them. Probably hardly two lapidaries would work exactly alike in their treatment of precious stones, and each guards his secrets, yet all use approximately similar general methods. Some have devised mechanical holders which permit the repeated cutting of stones to exactly the same angles, and that, too, with an accurate knowledge of the angles used. These angles can be definitely altered for different materials, according to their refractive indices. Other lapidaries produce very fine results by purely hand methods.

These details have been gone into to give an idea of the methods of the lapidary and of the many variations in method. In general, however, the *slitting* or *cleaving*, the *rubbing down* to shape, the *smoothing out* of all scratches and the *facetting* and *polishing* are done somewhat similarly by all lapidaries.

Having now had a glimpse of the methods of the lapidaries, let us briefly consider what constitutes good "make" in stones other than diamond.

Good "Make" in Colored Stones. Brilliants, cut from materials having smaller refractive indices than diamond, (and this group includes nearly all stones other than diamonds) should have steeper back angles and higher tops than the best diamond brilliants have. A 35-degree top angle (the angle between the slope of the top and the plane of the girdle is called the top angle) and a 41-degree back angle being about ideal for diamond, other gem materials should have more nearly a 39-degree top angle and a 44-degree back angle to give the greatest possible brilliancy. However, in the case of colored gems such as ruby, sapphire, etc., where the value depends even more largely upon the color than upon the brilliancy, it is frequently necessary to cut the brilliant thicker or thinner than these proportions in order to deepen or to thin the color.

In general, the thicker a stone of a given spread the deeper the color will be. The color may also be deepened by giving to the stone a rounded contour, both above and below the girdle, and facetting it in steps instead of in the brilliant form. Increasing the number of steps also serves to slightly deepen the color, as a larger number of reflections is thus obtained within the material, the light thus has to travel a greater distance through the colored mass, and more of the light, of color other than that of the stone, is absorbed.

Improving Color by Proper Cutting. In addition to the color improvement that can be brought about by changing the shape of the cut stone there are a number of gem materials whose color varies very greatly in different directions, and this fact calls for skillful use in order to obtain the best possible results. Thus most tourmalines of deep color must be cut with the top or table, of the finished stone, on the *side* of the prismatic crystal rather than at right angles to the axis of the prism. If cut the latter way they would be much too dense in color. On the other hand, most blue sapphires should be cut *across* the prism axis rather than the way that tourmalines should be cut. To cut a sapphire with its table on the side of the prism would be likely to cause it to have a greenish cast because of the admixture of the unpleasing "ordinary ray" of yellowish tint with the blue of the stone as seen up and down the prism. Some Australian sapphires are of a pronounced green when viewed across the axis of the crystal.

Rubies if cut, as was recommended for sapphires, give a very pure and very deep red color, but lack somewhat in the display of dichroism given by rubies that are cut with the table on the side of the crystal and parallel to its axis. Lapidaries need to know and to make use of such optical relations as these and jewelers might well inform themselves in such matters, especially if they have, or hope to acquire, trade in very fine colored stones.

Effect of Shape on Brilliancy. In actual practice it is common to find colored stones poorly cut for brilliancy, especially central brilliancy, and that, too, without the excuse of sacrifice of brilliancy in order to improve color. The fault is usually due to too great a desire to save size and weight. Frequently a stone would have greater value if properly cut, even at the expense of some size and weight. When stones are cut too shallow, as is frequently the case, they are sure to leak light in the center and they are thus weak and less brilliant there than they would be if made smaller in diameter and with steeper back slopes approximating 44 degrees.

Round stones, if their angles are correct, are more brilliant than stones of other contour such as square or cushion shape, or navette or heart shape. It can readily be seen that such odd shaped stones can hardly have the same top and back angles at every part of their circumference. If the angle from a corner of a square stone is correct then the angle from the middle of one side is obviously a little different. Small differences of angle make considerable differences in the brilliancy of cut stones. The prevailing tendency to cut nearly all diamonds round depends largely upon the above facts. In the case of colored stones, however, the added attractiveness which comes with odd or different contour more than makes up for the slight loss of brilliancy that may attend upon the shape selected. Such shapes as lend themselves to special designs in mountings also justify any little loss in brilliancy that accompanies the change in shape, provided the proportions retained give a considerable amount of total reflection within the stone and thus light up most of the stone as seen from the front.

The test of the "make" of a color stone is its appearance. If it lights up well over most of its surface and if the color is right, one should not criticize the "make" as one would be justified in doing in the case of a diamond. If, however, the effect is less attractive it would many times be advisable to measure the angles of the stone, or its thickness and spread as compared with similar measurements on a stone of fine appearance. Frequently one will thus find the reason for the failure of the stone to perform as it might, and recutting should be resorted to in such cases in order to get a smaller but more beautiful and hence more valuable stone.

LESSON XXIV

FORMS GIVEN TO PRECIOUS STONES

WHILE precious stones are cut to many different forms, there are, nevertheless, but a few general types of cutting. These may be classified as follows: First, the "*cabochon*" (Fig. 11) type of cutting; second, the old "*rose*" (Fig. 12) type of cutting; third, the *brilliant* (Fig. 13); fourth, the *step cutting* (Fig. 14).

Cabochons. Of these the first, or *cabochon* cutting, is probably the most ancient. The term comes from a French word signifying a bald pate (caboch, from Latin cabo, a head). The usual round cabochon cut closely resembles the top of a head in shape. Cabochon cut stones usually have a flat base, but sometimes a slightly convex base is used, especially in opals and in moonstones, and some stones of very dense color are cut with a concave base to thin them and thus to reduce their color. The contour of the base may be round, or oval, or square, or cushion shape, or heart shape or of any regular form. The top is always smooth and

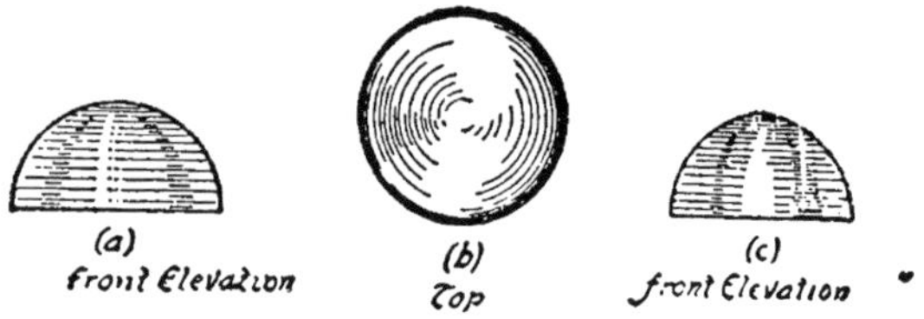

FIG. 11.—CABOCHON CUTTING.

rounding and unfacetted. The relation of the height or thickness to the length or width may be varied to suit the size and shape of the rough piece or to suit one's ideas of symmetry, provided the material be an opaque one, such as turquoise or lapis lazuli. If, however, the material is transparent the best results in the way of the return of light to the front, and hence in the display of the color of the material, are had if the thickness is about one half the spread.

This relation depends upon the refrae-tive index of the material, but as most color stones are of somewhat similar refractive indices, the above proportions are sufficiently accurate for all. The object in view is the securing of total reflection of as much light as possible from the flat polished back of the stone. Cabo-chon stones are sometimes set over foil or on polished gold to increase the reflection of light.

The path of a ray of light through a cabochon cut stone is closely similar to that through a rose cut diamond [see cut (c) of Fig. 12 for the latter.] Like the rose cut, the cabochon cut does not give much brilliancy as compared to the bril-liant cut. Cabochon cut stones, however, have a quiet beauty of color which commends them to people of quiet taste, and even fine rubies, sapphires, and emeralds are increasingly cut cabochon to satisfy the growing demand for fine taste in jewels. The East Indian has all along preferred the cabochon cut for color stones, but possibly his motives have not been unmixed, as the cabochon cut saves a greater proportion of the weight of the rough stone than the more modern types of cutting.

Garnets, more than other stones, have been used in the cabochon cut, and when

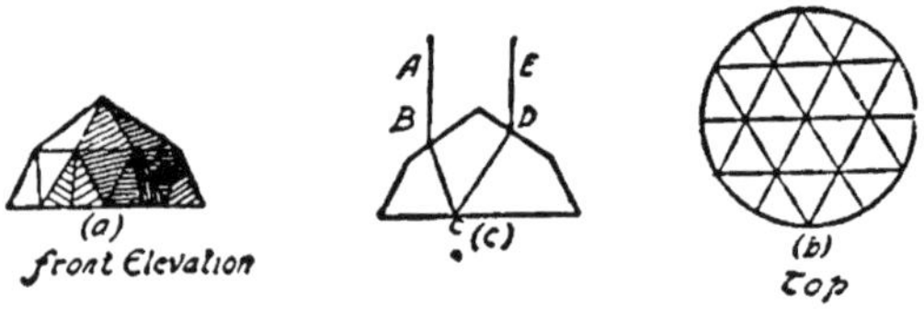

FIG. 12—ROSE CUTTING.

in that form are usually known as *carbuncles* (from carbunculus, a glowing coal). Any other fiery red stone might equally well be styled a carbuncle, especially if cabochon cut.

Scientific rubies look very well in the cabochon cut.

Fig. 11 shows in (a) and (b) the front and top of the usual round cabochon. Cut (c) of the same figure gives the front elevation of a cabochon which will light up better than the usual round topped design. In the round topped type the central part of the top is so nearly parallel to the back that light can pass right through as through a window pane. If the sloping sides are brought up to a blunt point, as in cut (c) there is very much less loss of light

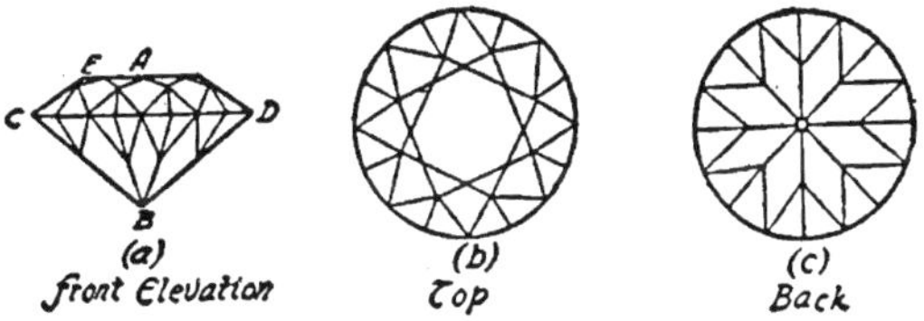

FIG. 13—BRILLIANT CUTTING

and greater beauty results. The East Indian cabochons are frequently cut in a fashion resembling that suggested.

Rose Cut Stones. It was natural that the earliest cut stones should have the simple rounded lines of the cabochon cutting, for the first thing that would occur to the primitive worker who aspired to improve upon nature's product, would be the rubbing down of sharp edges and the polishing of the whole surface of the stone. Perhaps the next improvement was the polishing of flat facets upon the rounded top of a cabochon stone. This process gives us the ancient type of cutting known as the *rose* cut. The drawings (a) and (b) of Fig. 12 show the front elevation and the top and (c) shows the path of a ray of light through a "rose." It will be noted that the general shape resembles that of a round cabochon, but twenty-four triangular facets have been formed upon the top. The well-proportioned rose has a thickness about one half as great as its diameter. Diamonds were formerly cut chiefly in the rose form, especially in the days of the East Indian mines, and even in the early part of the nineteenth century many people preferred finely made roses to the thick, clumsy brilliants of that day. To-day only very small pieces of diamond are cut to "roses." As the material so used frequently results from the cleaving of larger diamonds, the public has come to know these tiny roses as "chips."

The best roses have twenty-four regular facets but small ones frequently receive only twelve, and those are seldom regular in shape and in arrangement. Such roses serve well enough for encrusting watch cases and for similar work, as the flat base of the stone can be set in thin

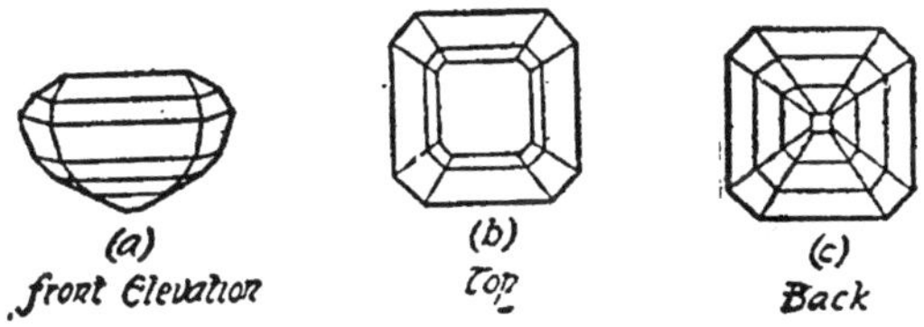

FIG. 14—STEP CUTTING.

metal without difficulty. About the only gem other than diamond that is now cut to the rose form is garnet. Large numbers of small Bohemian garnets are cut to crude rose form for use in cluster work.

The brilliant cut, as its name implies, gives the most complete return of light of any of the forms of cutting. The theory of the brilliant has already been discussed (Lesson XXII. in connection with the cutting of diamond). The shape of the brilliant is too well known to require much description. Most brilliants to-day are cut practically round and the form is that of two truncated cones placed base to base. The upper cone is truncated more than the lower, thus forming the large, flat top facet known as the *table* of the stone [A, Fig. 13, cut (a)]. The truncating of the lower cone forms the tiny facet known as the culet, which lies opposite to the table and is parallel to the latter [see B, Fig. 13, cut (a)]. The edge of meeting of the two cones is the *girdle* of the brilliant [CD in cut (a), Fig. 13]. The sloping surface of the upper cone is facetted with thirty-two facets in the full cut brilliant, while the lower cone receives twenty-four.

Small stones sometimes receive fewer facets, to lessen the cost and difficulty of cutting, but by paying sufficient for them full cut brilliants as small as one hundred to the carat may be had. Cut (b) of Fig. 13 shows the proper arrangement of the top facets and cut (c) that of the bottom facets.

When cutting colored stones in the brilliant cut, especially if the material is very costly and its color in need of being darkened or lightened, the lapidary frequently takes liberties with the regular arrangement and proportions depicted in the cuts.

Step Cutting. The only remaining type of cutting that is in very general use is the *step cut* (sometimes known as trap cut). Fig. 14, (a), (b), and (c), shows the front elevation, the top and the back of a square antique step cut stone. The contour may be round or completely square or oblong or of some other shape, just as a brilliant may have any of these contours. The distinctive feature of the step cutting is the several series of parallel-edged quadrangular facets above and below the girdle and the generally rounding character of its cross section. This plump, rounding character permits the saving of weight of the rough material, and by massing the color gives usually a greater depth of color than a brilliant of the same spread would have if cut from similar material. While probably never quite as snappy and brilliant as the regular brilliant cut, a well-proportioned step cut stone can be very brilliant. Many fine diamonds have recently been cut in steps for use in exclusive jewelry.

The Mixed Cut. The ruby and the emerald are never better in color than when in the full step cut, although rubies are frequently cut in what is known as the *mixed* cut, consisting of a brilliant cut top and a step cut back. Sapphires and many other colored stones are commonly cut in the mixed cut. Recently it has become common to polish the tops of colored stones with a smooth unfacetted, slightly convex surface, the back being facetted in either the brilliant or the step arrangement. Such stones are said to have a "*buffed top.*" They are less expensive to cut than fully facetted stones and do not have the snappy brilliancy of the latter. They do, however, show off the intrinsic color of the material very well.

LESSON XXV

IMITATIONS OF PRECIOUS STONES

"Paste" Gems. Large volumes have been written on paste jewels, especially on antique pastes. Contrary to a prevailing belief, the paste gem is not a recent invention. People frequently say when told that their gems are false, "But it is a very old piece, it must be genuine." The great age of a jewel should rather lead to suspicion that it was not genuine than give confidence that a true gem was assured. The Egyptians and Romans were skillful makers of glass of the sort used in imitating gems and some of the old pastes were very hard or else have become so with age.

Glass of one variety or another makes the most convincing sort of imitation precious stones. The term "paste" as applied to glass imitations is said to come from the Italian *pasta* meaning dough, and it suggests the softness of the material. Most pastes are mainly lead glass. As we saw in Lesson XVIII., on the chemical composition of the gems, many of them are silicates of metals. Now glasses are also silicates of various metals, but unlike gem minerals the glasses are not crystalline but rather amorphous, that is, without definite geometric form or definite internal arrangement.

The optical properties of the various glasses vary chiefly with their densities, and the denser the material the higher the refractive index and the greater the dispersion. Thus to get the best results in imitation stones they should be made of very heavy glass. The dense flint glass (chiefly a silicate of potassium and lead) which is used for cut glass ware illustrates admirably the optical properties of the heavy glasses. By using even more lead a still denser glass may be had, with even a greater brilliancy.

Unfortunately the addition of lead or other

heavy metals (such as thallium) makes the product very soft and also very subject to attack by gases such as are always present in the atmosphere of cities. This softness causes the stones to scratch readily so that when worn they soon lose their polish and with the loss of polish they lose their beauty. The attack of the gases before mentioned darkens the surfaces of the imitation and further dulls it. When fresh and new a well cut piece of colorless paste has a snap and fire that approaches that of diamond. The surface luster is not adamantinc, however, and the edges of the facets cannot be polished so sharply as those on a diamond. Moreover the refractive index, while high, is never so high as in a diamond and hence the brilliant cannot be so shaped as to secure the amount of total reflection given by a well-made diamond. Hence, the paste brilliant, while quite satisfying as seen from squarely in front, is weak and dark in the center as seen when tilted to one side. By these differences the trained eye can detect paste imitations of diamond at a glance without recourse to tests of specific gravity, hardness, etc.

Pastes, being amorphous, are singly refracting, as is diamond. This fact helps the appearance of the paste brilliant, for light does not divide within it to become weakened in power. This singleness of refraction, however, betrays the paste imitation when it is colored to resemble ruby, sapphire or emerald, all of which are doubly refracting.

The color is imparted to pastes by the addition, during their manufacture, of various metallic oxides in small proportions. Thus cobalt gives a blue color, copper or chromium green, copper or gold give red (under proper treatment) and manganese gives purple. By experiment the makers of pastes have become very skillful in imitating the color of almost any precious stone. Fine paste emeralds may look better than inferior genuine emeralds.

As pastes are singly refracting and hence

lack dichroism, the pleasing variety of color of the true ruby cannot be had in a paste imitation, but the public is not critical enough to notice this lack. The expert would, however, note it and could detect the imitation by that difference as well as by the lack of double refraction. The use of direct sunlight and a white card as already explained in the lesson on double refraction (Lesson III.) will serve to expose the singleness of refraction of paste imitations. Spinels and garnets are about the only true gems (except diamond) that are single refracting. Any other color stone should show double refraction when tested by the sun-light-card method. The file test will also expose any paste imitation as all the very brilliant pastes are fairly soft.

Doublets. To give better wearing quality to paste imitations the *doublet* was devised. This name is used because the product is in two parts, a lower or back portion of paste and an upper or top portion of some cheap but hard genuine stone. Garnet is probably used for this purpose to a greater extent than any other material, although quartz or colorless topaz will do very well.

The usual arrangement of the parts can be seen in Fig. 15, the garnet covering only

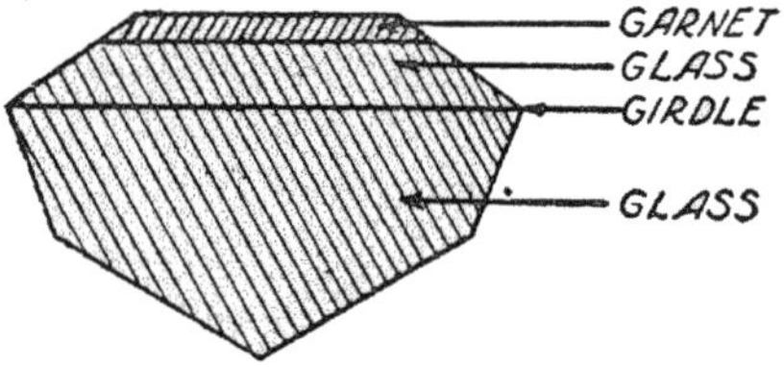

FIG. 15. ONE FORM OF CHEAP DOUBLET.

a part of the upper surface, namely the table part and a small portion of the sloping surface of the top. In high class doublets the hard mineral covers the paste to the girdle. (See Fig. 16.) The color of the garnet does not interfere seriously with that of the paste.

If a "diamond" doublet is desired the slice of garnet is made nearly as thin as paper and it covers only the table of the brilliant. It is thus practically colorless. A thin slice of red garnet over a green background is not noticeable, as all the red is absorbed in passing through the green material beneath. With a blue base, the red upper layer may give a very slight purple effect. With yellow a slight orange tint results and of course with

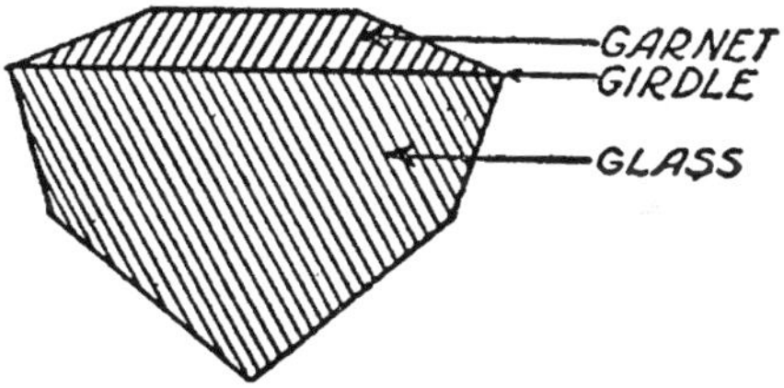

FIG. 16. ANOTHER FORM OF DOUBLET.

a red back no perceptible difference would result.

The two materials are cemented together, by means of a transparent waterproof cement. The *triplet* has already been described in Lesson XII. It is even better than the doublet and more difficult to detect. Both the file test and the sunlight-card test serve to detect doublets, as well as paste imitations, except that in the file test with the fully protected doublet the *back* of the stone must be tested with the file, as the girdle and top are of hard material.

In the sunlight-card test of a doublet (the refraction of garnet being single like that of glass), single images of the facets will be had on the card when the sunlight is reflected onto it. A reflection of the lower or inner surface of the garnet top can be seen also and this serves to still further identify a doublet or a triplet. The appearance of this reflection is much like that received on the card from the top of the table. It is larger than the reflections of the smaller facets and is but little colored.

Tests for Doublets. A trained eye can also detect a doublet or a triplet by noting the difference in the character of the surface luster of the garnet part and of the glass part. Garnet takes a keener and more resinous luster than glass. By tipping the doublet so that light is reflected to the eye from the sloping top surface, one can see at once where the garnet leaves off and the glass begins. Even through a show window one can tell a doublet in this way although here it is necessary to move oneself, instead of the stone, until a proper position is obtained to get a reflection from the top slope of the doublet.

If the garnet covers the whole top of the imitation then it is not possible to get so direct a comparison, but even here one can look first at the top surface and then at the back and thus compare the luster. It is also well to closely examine with a lens the region of the girdle, to see if any evidence of the joining of two materials can be seen. Frequently the lapidary bevels the edge so as to bring the line of junction between real and false material at the sharp edge of the bevel. Boiling a doublet in alcohol or chloroform will frequently dissolve the cement and separate the parts.

The dichroscope also serves to detect the false character of doublets and paste imitations, as neither shows dichroism. As rubies, emeralds, sapphires, and in fact most colored stones of value, show distinct dichroism, this test is a sure one against these imitations.

Triplets and doublets too may be exposed by dipping them *sidewise* into oil, thus removing the prismatic refraction almost completely, as the oil has about the same refractive index as the stone. One can then look directly through glass and garnet, or other topping material, separately, and each material then shows its proper color. Thus zones of color appear in a doublet or triplet when under the oil. A real gem would appear almost uniform in color under these conditions.

Round gas bubbles can frequently be found in paste, and hence in the paste part of a doublet. Also, the natural flaws of the real stone are never found in paste, but may be present in the real stone part of a doublet or a triplet. Some imitation emeralds on the market, however, have been made in a way to counterfeit the flaws and faults generally found in this stone.

Altered Stones. In addition to the out and out imitations made of paste, and the doublets, there are numerous imitations current in the trade that are made by staining or by otherwise altering the color of some genuine but inexpensive gem material.

For example, large quantities of somewhat porous chalcedony from Brazil are stained and sold in imitation of natural agate or sard or other stones. In many cases the staining is superficial, so that the stone has to be shaped before it is stained, then stained and polished.

Large quantities of slightly crackled quartz are stained to resemble lapis lazuli, and sold, usually with the title "Swiss Lapis." A file test will reveal the character of this imitation, as it is harder than a file, while true lapis is softer. The color too is never of so fine a blue as that of fine lapis. It has a Prussian blue effect.

Turquoises of inferior color are also sometimes stained to improve them. A better product is made artificially.

Opals are sometimes impregnated with organic matter, which is then charred, perhaps with sulphuric acid, thus giving them somewhat the appearance of black opal.

Opals are also imitated by adding oxide of tin to glass, thus imparting a slight milkiness to it. The imitation is then shaped from this glass by molding, and the back of the cabochon is given an irregular surface, which may be set over tinsel to give the effect of "fire."

Pale stones are frequently mounted over foil, or in enameled or stained settings and thus their color is seemingly improved.

Diamonds of poor color are occasionally "painted"; often the back of the brilliant is treated with a violet dyestuff, which even in so small an amount that it is difficult to detect, will neutralize the yellow of the stone and make it appear to be of a fine blue-white color. The "painting" is, of course, not permanent, so that such treatment of a diamond with a view to selling it is fraudulent. The painted stone may be detected by washing it with alcohol, when the dye will be removed and the off-color will become apparent. If the stone is unset one can see with a lens a wavery metallic appearance on the surfaces that have been "painted." This effect is due to the action of the very thin film of dye upon the light that falls upon it.

Besides the staining of genuine materials, they are sometimes altered in color by heat treatment, and this topic will be discussed in the next lesson.

LESSON XXVI

ALTERATION OF THE COLOR OF PRECIOUS STONES

MANY gem minerals change color when more or less strongly heated. Extreme heat whitens many colored materials completely.

"**Pinked Topaz.**" John Ruskin advises us to "seek out and cast aside all manner of false or dyed or altered stones" but, in spite of his advice, perhaps the most justifiable use of heat treatment is that which alters the color of true topaz from a wine-yellow to a fine pink. It would appear that the wine-yellow is a composite color composed of pink and yellow and that the pink constituent is less easily changed by heat than is the yellow one. If too high a temperature is used both colors disappear and white topaz results. As the latter is abundant in nature and of little value, such a result is very undesirable. Pink topaz, however, is very rare, and until recently, when pink tourmaline from California and Madagascar, and pink beryl (morganite) from Madagascar, became available in quantity, the "pinked" topazes had but few competing gems, and thus commanded a higher price than the natural topazes. Of course, care has to be taken in heating a mineral to raise and lower the temperature slowly, in order to avoid sudden and unequal expansion or contraction, which would crack and ruin the specimen, as the writer learned to his sorrow with the first topaz that he tried to "pink."

Spanish Topaz. Another material that gains a more valuable color by heat treatment is the smoky quartz of Spain, which, on being gently heated, yields the so-called Spanish topaz. Some amethysts are altered to a yellow color by mild heating. Too great a temperature completely decolorizes colored quartz. Some dark quartz yields a nearly garnet red product, after heating.

Zircon. Slight increase in temperature causes many of the zircons from Ceylon to change markedly in color. An alcohol flame serves admirably to effect the change, care being taken to warm up the stone very gradually and to cool it slowly. Drafts should be prevented, as they might suddenly cool the stone and crack it. Some zircons become completely whitened by this treatment. At the same time they increase markedly in density and in refractive index and thus become even more snappy and brilliant than when colored. One is tempted to suspect that the "space lattice" of the crystal has had its strata drawn closer together during the heating and left permanently in a closer order of arrangement. Other zircons merely become lighter colored and less attractive. Some of the whitened stones again become more or less colored on exposure to strong light. Ultra-violet light will sometimes restore these to a fine deep color in a short time.

The whitened zircon, when finely cut in the brilliant form, with truly flat facets and sharp edges and with a top angle of about 39 degrees and a back angle of about 44 degrees, so closely resembles a diamond that it will deceive almost anyone on casual inspection. The expert, even, may be deceived, if caught off his guard. The writer has a fine specimen of a little over one carat, with which he has deceived many jewelers and pawnbrokers, and even an importer or two. If it is presented as a stone that closely resembles diamond your expert will say: "Yes, it is pretty good, but it would never fool me." If, however, you catch him off his guard by suggesting, perhaps, "Did you ever see a diamond with a polished girdle?", then he will look at it with interest, remark on its fine color and "make," and never think of challenging its character.

The refractive index of the dense type

of zircon is so high (1.92–1.98) that it lights up well over most of the surface of the brilliant when cut, as above indicated, and does not show markedly the weak dark center shown by white sapphire, white topaz, colorless quartz, colorless beryl, and paste, when seen from the side. Moreover, the luster of zircon is nearly adamantine, so the expert does not miss the cold metallic glitter as he would with any other white stone. The color dispersion, too, is so high (86% as great as in diamond) that the zircon has considerable "fire," and thus the casual handler is again deceived. A fine white zircon is really prettier than a *poor* diamond. It cannot compare, however, with a *fine* diamond. It would never do to let an expert see your zircon beside even a fair diamond. The zircon would look "sleepy." It is only when no direct comparison is possible, and when the expert is not suspicious, that a zircon can deceive him. Of course, the use of the scientific tests of the earlier lessons will, at once, detect the character of a whitened zircon. The hardness is but 7.5, the refraction so strongly double that the edges of the back facets appear double-lined when viewed through the table with a lens, and the specific gravity is 4.69. Double spots of light appear on the card when the sunlight-card test is applied. Hence, it is easy to detect zircon by any of these tests if there is reason to suspect that it has been substituted for diamond.

Corundum Gems. Rubies of streaky color are said to be improved by careful heating. Usually ruby undergoes a series of color changes on being heated, but returns through the same series in reverse order on being cooled, and finally resumes its original color. Strong heating will whiten some yellow sapphire. The author thus obtained a white sapphire from a crystal of light yellow material.

It is interesting to note that the corundum gems undergo marked change in color under the influence of radium. A regular series of changes is said to be produced in white sapphire by this means, the final color being yellow. This color may then be removed by heat and the series run through again. It is not stated that a fine red has ever been thus obtained. Perhaps Nature, by her slower methods, using the faint traces of radio-active material in the rocks, reddens the corundum of Burmah at her leisure, and finally arrives at the much sought "pigeon blood" color. It is said that the natives of India have a legend to the effect that the white sapphires of the mines are "ripening rubies," and that one day they will mature. Perhaps they are not far wrong.

Diamond. Diamonds of yellowish tint may be improved in color by the use of high-power radium. At present the latter is so rare and costly that there is no evidence of its commercial use for this purpose. Scientists have brought about the change to a light blue as an experiment. It is not yet known whether the change will be permanent. Perhaps here again Nature has anticipated man's discovery and made the fine bluish-violet Brazilian diamonds (which fluoresce to a deep violet under an arc light, and which shine for a few moments in the dark after exposure to light) by associating them for ages with radio-active material. Some of the African stones also have these characteristics.

Aside from the change in the color of diamond that may be brought about by means of radium, the mineral is extremely reluctant to alter its color. Many experimenters besides the author have tried in vain a host of expedients in the hope of finding some way to improve the color of diamond. About the only noticeable alteration that the author has been able to bring about was upon a brown diamond, the color of which was made somewhat lighter and more ashen by heating it in a current of hydrogen gas to a low red heat.

LESSON XXVII

PEARLS

UNLIKE the gems that have been so far considered, the pearl is not a mineral, but is of organic origin, that is, it is the product of a living organism. There are two principal types of molluscs which yield true pearls in commercial quantities. The best known of the first type is the so-called pearl oyster (*meleagrina margaritifera*). The pearl mussel of fresh water streams is of the second type (*uniomargarifer*). Other species of molluscs having pearly linings to their shells may produce pearls, but most of the pearls of commerce come from one or the other of the two varieties mentioned.

Structure of Pearl. The structure and material of the true pearl must be first understood in order to understand the underlying reasons for the remarkable beauty of this gem. Pearls are composed partly of the mineral substance calcium carbonate (chemically the same as marble) and partly of a tough, horny substance of organic nature called conchiolin. The shell of the pearl-bearing mollusc is also composed of these two substances. Calcium carbonate may crystallize in either of two forms, calcite or aragonite. In marble we have calcite. In the outer portions of the shell of the pearl oyster the calcium carbonate is in the form of calcite, but in the inner nacreous lining and in the pearl itself the mineral is present as aragonite. This is deposited by the mollusc in very thin crystalline layers in the horny layers of conchiolin, so that the lining of the shell is built of approximately parallel layers of mineral and of animal substance. In the normal shell this is all that takes place, but in the case of a mollusc whose interior is invaded by any small source of irritation, such as a borer, or a grain of sand, or other bit of foreign material, a process of alternate deposit of conchiolin and of aragonite goes on upon the invading matter, thus forming a pearl.

The pearl is built in layers like an onion. In shape it may be spherical, or pear shaped, or button shaped or of any less regular shape than these. The regular shapes are more highly valued. The spherical shape is of greatest value, other things being equal. Next comes the drop or pear shape, then the button shape, and after these the host of irregular shapes known to the jeweler as "baroques." The river man who gathers mussels calls these odd shaped pearls "slugs."

Let us now attempt to understand how the beautiful luster and iridescence of the pearl are related to the layer-like structure of the gem. In the first place, it should be understood that both conchiolin and aragonite are translucent, that is, they pass light to a certain extent. The layers being exceedingly thin, light can penetrate a considerable number of them if not otherwise deflected from its course. We thus obtain reflections not merely from the outer surface of a pearl, but from layer after layer within the gem and all these reflections reach the eye in a blended reflection of great beauty. The luster of a pearl is then not purely a *surface luster* in the usual sense of that term, but it is a luster due to many superposed surfaces. It is so different from other types of luster that we describe it merely as *pearly luster* even though we find it in some other material, as, for example in certain sapphires, in which it is due to a similar layer-like arrangement of structure.

Orient. The fineness of the luster of a pearl, or as is said in the trade, the *orient*, depends upon the number of layers that take part in the reflection, and this number in turn depends upon the translucency of the material and the thinness of the layers. Very fine pearls usually have very many, very thin layers taking part in the reflection. The degree of translucency, considered apart, is sometimes called the "water" of the pearl.

In addition to their beautiful luster, many pearls display iridescence, and this is due in part, as in the case of the pearly lining of the shell (mother of pearl) to overlapping of successive layers, like the overlapping of shingles on a roof. This gives rise to a lined surface, much like the diffraction grating of the physicist, which is made by ruling a glass plate with thousands of parallel lines to the inch. Such a grating produces wonderful spectra, in which the rainbow colors are widely separated and very vivid. The principal on which this separation of light depends is known as diffraction and cannot be explained here, but a similar effect takes place when light falls on the naturally ruled surface of a pearl and helps produce the play of colors known as iridescence. The thin layers themselves also help to produce the iridescence by interference of light much as in the case of the opal, which has already been discussed.

Color. Having explained the cause of the orient and water of pearls, the *color* must next be considered. Pearls may be had of almost any color, but the majority of fine pearls are white, or nearly so. The fine Oriental pearls frequently have a creamy tint. Among fresh water pearls the creamy tint is less often seen, but fine pink tints occur. Occasionally a black pearl is found and on account of its rarity commands a price nearly as great as that obtainable for a white pearl of similar size and quality.

The value of pearls depends upon several different factors and it is far from an easy matter to estimate the value of a fine specimen. It is much easier to grade and estimate the value of diamonds than to do the same for pearls, and it is only by long and intimate acquaintance with the pearls themselves that one can hope to become expert in deciding values. There are, however, several general factors that govern the value of pearls. Chief among these are: 1, *Orient;* 2, *Color;* 3, *Texture or Skin;* 4, *Shape and Size.*

Factors Governing the Value of Pearls. Taking up each of these factors in turn, it may be said of the first that unless a pearl has that fine keen luster known as a fine orient, it is of but limited value. No matter what the size, or how perfect the shape, it is nothing, if dead and lusterless. To have great value the gem must gleam with that soft but lively luster peculiar to fine specimens of pearl. With variations in orient go wide variations in value.

As to *color*, the choicest pearls are pure white or delicate rose pink or creamy white. Pearls in these shades can be had in numbers and these colors are what might be called *regular* colors. *Fancy colored* pearls have peculiar and irregular values, depending a good deal upon rarity and upon the obtaining of a customer for an odd color. Fine pink and fine black pearls are examples of the type that is meant here.

To be very valuable a pearl must have a smooth even *skin*, that is, the *texture* of its surface must be even and regular. It must not have pits or scratches or wrinkles, or little raised spots upon it, or any cracks in it. In connection with this topic of "skin," it may be mentioned that it is sometimes true that a pearl of bad skin or of poor luster may be improved markedly by "peeling" it, as the process is called. As was said above, a pearl is built in layers much like an onion, and it can often be peeled, that is, one or more layers can be removed, thus exposing fresh layers beneath, whose texture and luster may be better than those of the original outside layer.

"Peeling" a Pearl. Possibly an anecdote of an actual case may serve best to explain the method by which "peeling" is sometimes accomplished. The writer was once at Vincennes, Ind., on business, and there became acquainted with a pearl buyer who was stopping at that place to buy fresh water pearls and "slugs" from the rivermen who gather the mussels for the sake of their shells. The latter are made into "pearl" buttons for clothing. It happened that the pearl buyer had accumulated some twenty-eight ounces of slugs and a number of pearls and was leaving on the same train with the author, who shared his seat with him. While we were looking over the slugs together the pearl buyer put his hand in his pocket and drew out a five dollar bill which he unrolled, exposing a pearl of about six grains, well shaped, but of rather dead luster. Remarking that he had paid but $4 for it and that he had rolled it up in the bill for safe keeping until he got time to peel it, he took out a small penknife, opened one of the blades, put a couple of kid glove finger tips on the thumb and first finger of his left hand and proceeded to peel the pearl on the moving train. Holding his two hands together to steady them, he pressed the edge of his knife blade against the pearl until the harder steel had penetrated straight down through one layer. Then with a flaking, lateral motion he flaked off a part of the outer skin. Bit by bit all of the outer layer was flaked off, and that, too, without appreciably scratching the next layer, so great was the worker's skill. When the pearl was completely peeled it was gently rubbed with three grades of polishing paper, each finer than the previous one, and then the writer was allowed to examine it. The appearance had been much improved, although it was not of extremely fine quality even when peeled. Under a high power magnifier scarcely a trace of the peeling could be seen. The value of the $4 pearl had been raised to at least $100 and not many minutes had been required for the change. A slower and more laborious, but safer, process of "peeling" a pearl, consists in gently rubbing the surface with a very fine, rather soft, abrasive powder until all of the outer skin has been thus worn away.

Of course, in many such cases no better

skin than the outer one could be found and disappointment would result from the peeling of such a pearl. It should be added that it will not do to try to peel a *part* of a pearl in order to remove an excrescence, for then one would inevitably cut across the layers, exposing their edges, and such a surface looks, when polished, much like a pearl button, but not like a pearl.

In this connection may be mentioned the widespread belief on the part of the public that the concretions found in the common edible oyster can be polished by a lapidary, as a rough precious stone can be improved by the latter, and that a fine pearl will result. It is frequently necessary for jewelers to whom such "pearls" are brought, to undeceive the person bringing them and to tell him that only those molluscs that have a beautiful pearly lining to their shells are capable of producing true pearls and that the latter require no assistance from the lapidary.

Shape. To return to the topic of factors governing the value of pearls, the *shape* of the pearl makes a vast difference in the value. Perfectly spherical pearls are most highly valued and closely following come those of drop or pear shape, as this shape lends itself nicely to the making of pendants. Oval or egg shaped pearls are also good. After these come the button shapes, in which one side is flattened. Pearls of irregular shape are much less highly valued. The irregular shaped pearls are called *baroque* pearls in the trade. The rivermen engaged in the fresh water pearl fishery call them *slugs*. Some of the more regular of these are called "nuggets." Others are termed "spikes" because of their pointed shape, and still others are called "wing" pearls on account of their resemblance to a bird's wing. Most of the baroques are too irregular in shape to have any special name applying to their form.

Weight. After orient, color, skin, and shape have been considered, *size* or *weight* finally determines the value. Pearls are sold by an arbitrary unit of weight known as the *pearl grain*. It is not equal to the grain avoirdupois, but is one fourth of a diamond carat. As the new metric carat is one fifth of a gram and as there are 15.43 avoirdupois grains in a gram, it is seen at once that there are but 3.08 real grains in a carat rather than four. Thus the *pearl grain* is slightly lighter than the avoirdupois grain.

Since large, fine pearls are exceedingly rare, the value mounts with size much more rapidly than is the case with any other gem; in fact, the value increases as the *square* of the weight. For example, let us consider two pearls, one of one grain weight, the other of two grains, and both of the same grade as to quality. If the smaller is worth say $2 per grain, then the larger is worth 2 × 2 (the square of the weight) times $2 (the *price per grain base*, as it is called in the trade), which totals $8. A four-grain pearl of this grade would be worth 4 × 4 × $2 = $32, etc. Thus it is seen that the price increases very rapidly with increase in weight.

Price "Per Grain Base." Some of the lower grades of pearls in small sizes are sold by the grain *straight*, that is, the price per grain is merely multiplied by the weight in grains to get the value, just as the price per carat would be multiplied by the number of carats to get the value of a diamond. This method of figuring the value of pearls is used only for the cheaper grades and small sizes, however, and the method first explained, the calculation per grain *base*, is the one in universal use for fine gems. Very fine exceptional gems may be sold at a large price *for the piece*, regardless of the weight.

It is interesting to note in this connection that Tavernier, the French gem merchant of the seventeenth century, tells us that in his day the price of large diamonds was calculated by a method similar to that which we now use for pearls, that is, the weight in carats was squared and the product multiplied by the price per carat. Such a method would give far too high a price for diamonds to-day.

The High Price of Fine Pearls. This suggests the thought that pearls of fine quality and great size are the most costly of all gems to-day and yet there seems to be no halting in the demand for them. In fact, America is only just beginning to get interested in pearls and is coming to esteem them as they have long been esteemed in the East and in Europe. Those who have thought that the advance in the prices of diamonds in recent years will soon put them at prohibitive rates should consider the enormous prices that have been obtained and are being obtained for fine pearls.

In order to facilitate the calculating of prices of pearls, tables have been computed and published giving the values of pearls of all sizes at different prices *per grain base*, and several times these tables have been outgrown, and new ones, running to higher values, have been made. The present tables run to $50 per grain base.

There is much justification for the high prices demanded and paid for large and fine pearls. Such gems are really exceedingly scarce. Those who, as boys, have opened hundreds of river mussels only to find a very few small, badly misshapen "slugs" will realize that it is only one mollusc in a very large number that contains a fine pearl. Moreover, like the bison and the wild pigeon, the pearl-bearing molluscs may be greatly diminished in numbers or even exterminated by the greed of man and his fearfully destructive methods of harvesting nature's productions. In fact, the fisheries have been dwindling in yield for some time, and most of the fine pearls that are marketed are *old* pearls, already drilled, from the treasuries of Eastern potentates, who have been forced by necessity to accept the high prices offered by the West for part of their treasures. In India, pearls have long been acceptable collateral for loans, and many fine gems have come on the market after failure of the owners to repay such loans.

Having considered the factors bearing on the value of pearls, we will next consider briefly their physical properties. The specific gravity is less definite than with minerals and varies between 2.65 and 2.70. It may be even higher for pink pearls.

Physical Properties. In hardness pearls also vary, ranging between 3½ and 4 on Mohs' scale. They are thus very soft and easily worn or scratched by hard usage. A case showing the rather rapid wearing away of pearls recently came to the attention of the writer. A pendant in the shape of a Latin cross had been made of round pearls which had been drilled and strung on two slender gold rods to form the cross. The pearls were free to rotate on the wires. After a period of some twenty or more years of wear the pearls had all become distinctly cylindrical in shape, the rubbing against the garments over which the pendant had been worn having been sufficient to grind away the soft material to that extent. The luster was still good, the pearls having virtually been "peeled" very slowly by abrasion.

Care of Pearls. This example suggests the great care that should be taken by owners of fine pearls to prevent undue rubbing or wear of these valuable but not extremely durable gems. They should be carefully wiped after being worn to remove dust and then put away in a tightly closed case.

Pearls should never be allowed to come in contact with any acid, not even weak acids like lemonade, or punch or vinegar, as, being largely calcium carbonate they are very easily acted upon by acids, and a mere touch with an acid might ruin the surface luster. Being partly organic in nature, pearls are not everlasting, but must eventually decay, as is shown by the powdery condition of very old pearls that have been found with mummies or in ancient ruins. The organic matter has yielded to bacterial attack and decayed, leaving only the powdery mineral matter behind. As heat and moisture are the conditions most conducive to the growth of bacteria, and hence to decay, it would follow that fine pearls should be kept in a dry cool place when not in use.

LESSON XXVIII

CULTURED PEARLS AND IMITATIONS OF PEARLS

Cultured Pearls. Like all very valuable gems, pearls have stimulated the ingenuity of man to attempt to make imitations that would pass for genuine. Perhaps the most ingenious, as well as the most natural looking product, is the "*cultured pearl.*" This is really natural pearl on much of its exterior, but artificial within and at the back. In order to bring about this result the Japanese, who originated the present commercial product, but who probably borrowed the original idea from the Chinese, call to their assistance the pearl oyster itself. The oysters are gently opened, small hemispherical disks of mother-of-pearl are introduced between shell and mantle and the oyster replanted. The foreign material is coated by the oyster with true pearly layers as usual, and after several years a sufficiently thick accumulation of pearly layers is thus deposited on the nucleus so that the oyster may be gathered and opened and the cultured pearl removed by sawing it out from the shell to which it has become attached. To the base is then neatly cemented a piece of mother-of-pearl to complete a nearly spherical shape, and the portions of the surface that have not been covered with true pearl are then polished. The product, when set in a proper pearl mounting, is quite convincing and really beautiful.

As the time during which the oyster is allowed to work upon the cultured pearl is doubtless far less than is required for the growth of a large natural pearl, the number of layers of true pearly material is considerably smaller than the number of layers that take part in the multiple reflections explained in the previous lesson, and hence the "orient" of the cultured pearl is never equal to that of a fine true pearl. It is frequently very good however, and for uses that do not demand exposure of the whole surface of the pearl, the cultured pearl supplies a substitute for genuine pearls of moderate quality and price. The back parts of the cultured pearl, being only polished mother-of-pearl, have the appearance of the ordinary pearl button, rather than that of true pearl.

Imitations of Pearls. Aside from these half artificial cultured pearls, the out and out imitations of pearls that have been most successfully sold are of two general types, first "*Roman pearls*," and, second, "*Indestructible pearls.*" The Roman pearls are made hollow and afterward wax filled, the Indestructible pearls have solid enamel bases. In both types the pearly appearance is obtained by lining the interior, or coating the exterior, with more or less numerous layers of what is known as "*nacre*" or some times as "*essence d'oriente.*" This is prepared from the scales of a small fish found in the North Sea and in Russia. The scales are removed and treated with certain solutions which remove the silvery powder from the scales. The "*nacre*" is then prepared from this powder. The fineness of the pearly effect becomes greater as the preparation ages, so very fine imitations are usually made from old "*nacre*." The effect is also better the larger the number of successive layers used. The artificial pearl thus resembles the true pearl in the physical causes for the beautiful effect.

In some cases the Roman pearl has a true iridescence which is produced by "burning" colors into the hollow enamel bead. Some of the indestructible pearls are made over beads of opalescent glass, thus imparting a finer effect to the finished product. While the cheaper grades of indestructible pearls have but three or four layers of nacre, some of the fine ones have as many as thirty or more. The earlier indestructible pearls were made with a coating material which was easily affected by heat, or by water, or by perspiration, as a gelatine-like sizing was included in it. The more recent product has a mineral binder which is not thus affected, so that the "pearls" are really about as durable as natural ones, and will at least last a lifetime if used with proper care.

Like fine natural pearls, the fine imitations should be wiped after use and carefully put away. They should also be restrung occasionally, as should real pearls both to prevent loss by the breaking of the string and because the string becomes soiled after a time, and this hurts the appearance of the jewel.

The "Roman" type of imitation will not stand much heat, as the wax core would melt and run out.

Testing Imitations of Pearls. As the making of imitations of pearls is mainly handwork and as many treatments are required for the best imitations, fairly high prices are demanded for these better products, and the appearance and permanency warrant such prices. The best imitation pearls are really very difficult of detection except by close examination. They will not, of course, stand inspection under a high magnification.

Artificial pearls may also be detected by their incorrect specific gravity, by their incorrect degree of hardness, and in the case of the hollow pearls by making a tiny ink spot upon the surface of the "pearl" and looking at it through a lens. A reflection of the spot from the *inside* surface of the bead will appear beside the spot itself if the pearl is of the Roman type.

The artificial pearls so far described are high class products. Some of the very cheap and poor imitations are merely solid, or hollow, glass or enamel beads which have been made slightly pearly, either by adding various materials to the glass or enamel when it was made, or by crudely coating the beads without or within with wax containing cheap "nacre."

LESSON XXIX

THE USE OF BALANCES, AND THE UNIT OF WEIGHT IN USE FOR PRECIOUS STONES

AS precious stones are almost always sold by weight, and as the value at stake is frequently very great, it is almost as necessary for a gem merchant, as it is for the chemist, to have delicate balances and to keep them in good order and to use them skillfully.

A general understanding of the unit of weight in use for precious stones and how it is related to other standard weights is also necessary to the gem dealer. We will therefore consider in this lesson the use and care of balances and the nature and relative value of the unit of weight for precious stones.

Delicate Balances Needed. As it is necessary, on account of their great value, to weigh some gems, such as diamonds, emeralds, rubies, etc., with accuracy to at least the one hundredth part of a carat (which is roughly in the neighborhood of 1/15,000 of an ounce avoirdupois), balances of very delicate and accurate construction are a necessary part of the equipment of every gem merchant. While portable balances of a fair degree of accuracy are to be had, the best and surest balances are substantially constructed and housed in glass cases, much as are those of the analytic chemist, which must do even finer weighing. The case protects the balance from dust and dirt and prevents the action of air currents during the weighing. The balance itself has very delicate knife edges, sometimes of agate, sometimes of hardened steel, and these knife edges rest, when in use, on a block of agate or steel, so that there is a minimum amount of friction. When not in use the balance beam and knife edges are lifted from the block and held firmly by a metal arm, or else, as is the case with some balances, the post supporting the block is lowered, leaving the beam and knife edges out of contact with it. The object of this separation is to prevent any rough contact between the knife edges and the block on which they rest. Advantage should always be taken of this device whenever any fairly heavy load is put on or taken off of either pan, as the sudden tipping of the beam might chip the knife edges if not supported. When the load is nearly balanced there may be no harm in carefully adding or removing small weights while the knife edges are resting on the block, but even then it is safer to lower the beam and pans. It should be needless to state that as level and rigid a support should be had for one's balance as circumstances permit.

Method of Use of Balances. Before using a balance one should see that the pans are clean, that the base of the balance is properly leveled (the better balances have a spirit level attached) and that the pans balance each other without load. When slightly out of balance the defect may be adjusted by *unscrewing* the little adjusting nut at the end of the beam that is too light, or by *screwing in* the nut at the opposite end. Having seen that the adjustment is perfect the pans should be lowered and the object to be weighed placed on the *left-hand pan* (because a right handed person will find it handier to handle his weights on the right-hand pan). One should next guess as nearly as possible the weight of the stone and place well back on the right-hand pan the weight that he thinks comes nearest to that of the stone. If the weight is too heavy the next lighter weight should replace it. Smaller weights should be added until a perfect balance is had, the small weights being neatly arranged in the order of their size, in order to more rapidly count them when the stone is balanced. This is the case when the pointer swings approximately equal distances to the right and to the left and there is then no need to wait for it to come to rest in the center.

It is well to count the weights as they lie on the pan (which is easily done if they have been arranged in descending order of size as suggested above) then write down the total, and on removing the weights count aloud as they are replaced in the box and note if the total checks that which was written down. It may seem unnecessary to be so careful in this matter, but it is better to be over-careful than to make a mistake where every hundredth of a carat may mean from one to five or six dollars or more. No dealer can afford to have a stone that he has sold prove to be lighter than he has stated it to be. One should be at least within one one-hundredth of a carat of the correct weight.

It should be unnecessary to add that accurate weights *should never be handled with the fingers.* Ivory tipped forceps are best for handling the weights. The forceps commonly used for handling diamonds will, in time, wear away the weights by scratching them so that they will weigh materially less. Unless the weights are of platinum or plated with gold, the perspiration of the hands would cause them to oxidize and gain in weight. It would be well to discard the smaller weights, which are most in use, every few years and obtain new and accurate ones. In case this is not done one should at least have the weights checked against others known to be of standard weight. Any chemist will have balances and weights far more accurate than the best in use for precious stones and will gladly check the weights of a gem dealer for a moderate fee.

To check the accuracy of your balance, change the stone and weights to opposite pans, in which case they should still balance.

One should never overload a balance, both because the balance might be injured and because the relative accuracy decreases as the load increases. If the weight of a parcel of stones heavier than the total of the weights provided with the balance is desired, the parcel should be divided and weighed in parts.

While many dealers neglect some of the precautions above suggested and somehow get along, yet it is safer to use care and to have correct technique in the handling of one's balances.

Having indicated a few of the refinements of method in weighing we will next consider the unit of weight in use for precious stones and see how it is related to other units of weight and in what manner it is subdivided.

The Unit of Weight for Precious Stones. The present unit for precious stones in the United States is the *metric carat.* Most of the more progressive countries have in recent years agreed upon the use of this unit. Its use in the United States became general July 1, 1913. It is by definition exactly one fifth of a *gram* (the unit of weight of the *Metric System* of weights and measures). Its relation to the *grain* is that there are 3.08+grains in the metric carat. The carat in use in this country up to a few years ago was about 2½% heavier than the present metric carat. It was equal to

.2053 grams instead of .2000 grams (1/5 gram). The carats of countries not using the metric carat vary considerably, but yet approximate the metric carat somewhat nearly.

Thus, that in use in Great Britain was .2053 g., in Amsterdam .2057 g., in Berlin .20544 g., in Lisbon .20575 g., and in Florence 0.1972 g. The latter was the only one that was under the metric carat. The change to the metric carat was desirable, as it unified the practice of weighing, which not only varied in different countries, but even in the same country. Thus there was no very exact agreement among the makers of diamond weights in the United States prior to the adoption of the metric carat. One man's carat was a bit heavier or lighter than another's. With a definite and simple relationship to the standard gram there is now no excuse for any variation in weights. The Bureau of Standards at Washington affords manufacturers every facility for standardizing their weights.

The Decimal System of Subdivision of the Carat. With the adoption of the metric carat the custom of expressing parts of a carat in common fractions whose denominators were powers of the number 2 (½, ¼, ⅛, $\frac{1}{16}$, $\frac{1}{32}$, $\frac{1}{64}$) was discarded as awkward and slow for computation and the decimal system of subdivision was adopted. Thus the metric carat is divided into tenths and one hundredths. It is customary, however, to sum up the one hundredths and express them as the total number of one hundredths and not to express them as tenths. Thus, a stone of 2.57 carats is said to weigh "two and fifty-seven hundredths carats." The decimal system of subdivision of the carat makes the figuring of values simpler where no tables are handy. Of course, new tables were at once prepared when the new carat was adopted and they afford a rapid means of ascertaining the value of a stone of any weight when the price per carat is known. Should it become necessary to convert the weight of a stone from its expression in the old system to that of the new, one need only get $1.02\frac{1}{2}\%$ of the old weight. (The old carat was approximately .205 g., while the new one is .200 g. Hence one old carat

is $\frac{.205}{.200} = \frac{.102\frac{1}{2}}{.100} = 102\frac{1}{2}\%$ of a new one.)

Method of Converting Weights. If the old weight has fractions these should first be changed to decimals for convenience. For example, suppose it is wished to change $2\frac{1}{4}\ \frac{1}{16}$ old carats to metric carats. $\frac{1}{4} = .25$ and $\frac{1}{16} = .0625$. Hence $2\frac{1}{4}\ \frac{1}{16} = 2.3125$. Now get $102\frac{1}{2}\%$ of this: $(2.3125 \times 1.025 = 2.37$ metric carats).

If, for any reason one should need to change from metric carats to old U. S. carats one should multiply by .9756

$$\left(\frac{.200\text{ g.}}{.205\text{ g.}} = .9756\right)$$

As was said in Lesson XXV. pearls are sold

by the *pearl grain*, which is arbitrarily fixed at ¼ of a carat. With the change to the metric carat the pearl grain was correspondingly changed and its weight is now ¼ of .200 g. = .05 g., as expressed in the metric system.

LESSON XXX

TARIFF LAWS ON PRECIOUS AND IMITATION STONES

SINCE it is necessary for a nation, as well as for an individual, to have an income, and since articles of luxury are more easily taxed than are those of necessity, the traffic in gems and their imitations has frequently been made a source of revenue to our government. Usually the per cent. charged as tariff has been comparatively low, especially upon very valuable gems, such as diamonds and pearls, for the reason that too high a tariff would tend to tempt unscrupulous dealers to smuggle such goods into the country without declaring them. When the margin of difference between the values, with and without the tariff, is kept small the temptation is but slight, when the danger of detection and the drastic nature of the usual punishment are taken into account. Rough stones have frequently been allowed to enter the country duty free because they were regarded as desirable raw materials which would afford employment to home industry.

The tariff laws of October 3, 1913, made, however, some sweeping changes in the policy of our government toward precious stones and as those laws are still in force (April 4, 1917) this lesson will attempt to set forth clearly the exact conditions under the present law.

Perhaps the paragraph of first importance to the trade is No. 357 which reads as follows.

"357. Diamonds and other precious stones, rough or uncut, and not advanced in condition or value from their natural state by cleaving, splitting, cutting, or other process, whether in their natural form or broken, and bort; any of the foregoing not set, and diamond dust, 10 per centum ad valorem; pearls and parts thereof, drilled or undrilled, but not set or strung; diamonds, coral, rubies, cameos, and other precious stones and semi-precious stones, cut but not set, and suitable for use in the manufacture of jewelry, 20 per centum ad valorem; imitation precious stones, including pearls and parts thereof, for use in the manufacture of jewelry, doublets, artificial, or so-called synthetic or reconstructed, pearls and parts thereof, rubies, or other precious stones, 20 per centum ad valorem."

It will be noticed that the chief changes over the previous law are first that which imposes a 10% duty on rough precious stones, which were formerly free of duty, and second the advance in the duty on cut diamonds and other cut stones from the former 10% to the present 20%.

This increase in the tariff was regarded as unwise by many conservative importers, as the temptation to defraud the government is made much greater than before. The change was even feared by honest dealers who were afraid that they could not successfully compete with dishonest importers who might smuggle gems into the country. In spite of a rather determined opposition the change was made and our most representative dealers have been making the best of the situation and have been doing all that they could to help prevent smuggling or at least reduce it to a minimum. Through their knowledge of the movements of diamond stocks and of prices they are able to detect any unduly large supply or any unwarranted lowness of price and thus to assist the government agents by directing investigation towards any dealer who seems to be enjoying immunity from the tariff.

The question of the status of Japanese cultured pearls has been settled as follows. Paragraph 357 (quoted above) is ruled to cover them and they are thus subject to a 20% ad valorem tax.

Carbonadoes—miners' diamonds—are free of duty, under paragraph 474. Crude minerals are also free of duty, paragraph 549. Paragraph 607 declares "Specimens of natural history and mineralogy" are free.

In case the owner is not prepared to pay the tax on imported mechandise the government holds the goods for a period of three years pending such payments.

In case an importer shows that imported merchandise was purchased at more than actual market value, he may deduct the difference at time of entry and pay duty only on the wholesale foreign market value, under Section III, paragraph 1.

On the other hand, if the examiner finds merchandise to be undervalued on the invoice, such merchandise is subject to additional penal duties, but in case of disagreement between the importer and the examiner as to the actual market value, appeal may be taken to the Customs Court.

Since the Philippine Islands are possessions of the United States, pearls from those islands may be admitted free of duty when the facts of their origin are certified to.

In the case of precious stones which had their origin in the United States, but which were exported and kept for a time abroad it has been ruled that such stones may be imported into the United States free of duty.

When precious or imitation precious stones are imported into the United States and subsequently mounted into jewelry which is then exported, the duty which was paid upon entry may be refunded less a deduction of 1%.

The author wishes to extend his thanks to Examiner W. B. Treadwell of New York, for his assistance in regard to the subject dealt with in this lesson.

BIBLIOGRAPHY

THE student of gems will, of course, want to read many books on the subject and the following brief bibliography will enable the beginner to select his reading wisely from the start. Much more complete bibliographies will be found in some of the books listed here, one which is notably complete to date of publication is contained in *Diamonds and Precious Stones*, by Harry Emanual, F.R.G.S., London, John Camden Hatten, 1867. This covers many languages.

The book which will probably be found most useful by those who have mastered this little text is the work by G. F. Herbert-Smith, to which frequent reference has been made at the close of many of our chapters. It is thoroughly scientific, yet understandable, and is very complete on the scientific side of the subject.

Gem Stones, G. F. Herbert-Smith, Jas. Pott & Co., N. Y.

For another work and one which contains information of trade character as well as scientific information about gems see *Precious Stones* by W. R. Cattelle, J. B. Lippincott & Co., Phila., or see *A Handbook of Precious Stones*, by M. D. Rothschild, G. P. Putnam's Sons, N. Y.

Gems and Gem Minerals, by Oliver Cummings Farrington, A. W. Mumford, publisher, Chicago, 1903, is another good general work on gems. Its color plates of rough gem minerals are especially good.

Those who are especially interested in the diamond should see *The Diamond* by W. R. Cattelle, The John Lane Co., N. Y., which gives a good account of its subject and is rich in commercial information, or *Diamonds: A Study of the Factors which Govern their Value*, by the present author, G. P. Putnam's Sons, N. Y., 1914.

Sir Wm. Crook's, the *Diamond*, Harper & Bros., N. Y., is very interesting, especially in its account of the author's visits to the S. African mines.

Students of pearls will find *The Book of the Pearl*, by Dr. Geo. F. Kunz and Dr. Chas. Stevenson, Century Co., N. Y., very complete. A smaller work, yet a good one, on pearls is *The Pearl* by W. R. Cattelle, J. B. Lippincott & Co., Phila., 1907. This book is strong on the commercial side.

An older work is *Pearls and Pearling* by D. Edwin Streeter, Geo. Bell & Co., London.

A work on gems and gem-cutting by a practical cutter is *The Gem Cutter's Craft*, by Leopold Claremont, Geo. Bell & Sons, London, but it should be said that very few trade secrets will be found exposed in the book.

On the subject of scientific precious stones *The Production and Identification of Artificial Precious Stones*, by Noel Heaton, B.Sc., F.C.S., read before the Royal Society of Arts, Apr. 26, 1911, is very fine. It may be had in the annual Report of the Smithsonian Institution for 1911, p. 217. It gives one of the best accounts to be had of the history of the artificial production of precious stones, especially of the corundum gems. It also contains a splendid account of how to distinguish scientific from natural gems.

Most students of gems will need to refer frequently to some good text-book of mineralogy. Although old, Dana's *Mineralogy* is still a standard work. A newer book and one of a more popular nature is L. P. Gratacap's *The Popular Guide to Minerals*, D. Van Nostrand & Co., N. Y.

Among larger and more expensive books on gems may be mentioned *Precious Stones*, by Dr. Max Bauer. This is an English translation of a German work which is a classic in its field. As it is now out of print in its English edition, a somewhat detailed account of its character may be of value to those who may be inclined to go to the effort to seek a copy at a public library or perhaps to purchase one through second-hand book stores.

A popular account of their characters, occurrence and applications, with an introduction to their determination, for mineralogists, lapidaries, jewellers, etc. with an appendix on pearls and coral, by Dr. Max Bauer, Privy Councillor, professor in the Union of Marburg. Translated from the German by L. J. Spencer, M.A. (Cantab.), F.G.S., assistant in the mineral department of the British Museum. With twenty plates and ninety-four figures in the text. London., Chas. Griffin & Co., Ltd.: Phila., J. B. Lippincott Co., 1904.

The book is a large one. xv + 627 pages and is divided into three parts with an appendix on pearls and coral.

Part I. deals with the general characters of precious stones.

1. Natural characters and occurrence.
2. Applications of Precious Stones.
3. Classification of Precious Stones. 106 pages.

Part II. Systematic Description of Precious Stones, Diamond, Corundum Gems, Spinel, etc. 450 pages.

Part III. Determination and Distinguishing of Precious Stones. 20 pages.

Appendix, 26 pages. Pearls and Coral.

Bauer is exhaustive in his descriptions of the more important precious stones and he also describes briefly very many little known and little used gem minerals.

On forms of cutting he is old fashioned.

First 68 pages given to explanation of characters used in identifying stones. Good.

On the Process of Cutting. Pages 79–87. Good account. More practical than most books give.

Careful accounts of occurrence of precious stones with maps.

Character of the occurrence of diamond in India, Brazil, and Africa, quite in detail.

The student who wishes to master the subject of gems cannot afford to neglect Bauer.

For those who read French, the latest, the most complete and thorough book on gems is Jean Escard's *Les Pierres Précieuses*, H. Dunod et E. Pinat, Paris, 1914.

It is a large and finely illustrated work.

The author has really outdone Bauer. The detail in regard to diamonds especially is very fine. Even the use of diamonds in mechanical ways is very completely gone into and also details in regard to cutting diamonds are very completely given. It is to be hoped that an English translation will soon become available.

Another large and thorough going work is Gardner F. Williams' *The Diamond Mines of South Africa*, MacMillan, N. Y.

Dr. Geo. F. Kunz's *Gems and Precious Stones of North America*, The Sci. Pub. Co., N. Y., 1890, 336 pages, 8 colored plates (excellent ones too), many engravings, is a very complete account of all published finds of precious stones in the United States, Canada, and Mexico, giving a popular description of their value, history, archeology, and of the collections in which they exist, also a chapter on pearls and on remarkable foreign gems owned in the United States. Many rare and little known semi-precious stones are described here. Dr. Kunz is also the author of several more recent gem books notably *The Magic of Jewels and Charms* and *The Curious Lore of Precious stones*, Lippincott, Phila.

Among books on engraved gems is the old *Hand Book of Gem Engraving* by C. W. King; Bell & Daldy, London, 1866, and one by Duffield Osborne; Henry Holt & Co., N. Y. Another book on this subject is *Engraved Gems* by Maxwell Somerville; Drexel Biddle, Phila.

For those who wish still further references the following older works will prove interesting.

Precious Stones, by W. R. Cattelle; Lippincott, Phila. *Precious Stones*, by W. Goodchild; D. Van Nostrand & Co. N. Y.

Julius Wodiska, of New York, has also written an interesting work on precious stones, *A Book of Precious Stones*, Putnams, 1907.

Still older works are *Precious Stones and Gems* by Edwin W. Streetor; Chapman & Hall, London, 1877. This is a book of 264 pages with nine illustrations. It contains much of value and was unsurpassed in its day. Its first-hand accounts of numerous important, even celebrated diamonds and other precious stones will always make it valuable to the student of gems.

Another book by the same author is *The Great Diamonds of the World;* Geo. Bell & Sons, London, 1882; 321 pages. Not illustrated. Its title adequately describes its contents. It is an excellent work. The author even traveled in India tracing the history of some of the famous diamonds that he describes.

Diamonds and Precious Stones, by Louis Dieulafait published in its English translation by Scribner, Armstrong & Co., N. Y., 1874, is another old but interesting work. It has 292 pages and 126 engravings on wood. It gives a fine account of diamond cutting as practiced at that time. There is also an excellent history of the production of artificial precious stones to that date.

The Natural History of Precious Stones and of the Precious Metals by C. W. King, M. A., Bell & Daldy, London, 1870, is rich in references to classical literature.

One or two interesting monographs on precious stones have been written and *The Tourmaline*, by Augustus C. Hamlin is one of these. Mr. Hamlin became interested in gems because of his accidental discovery of some of the fine tourmalines of Maine. His *Leisure Hours among the Gems* is also very readable. Jas. R. Osgood & Co., Boston, 1884. It deals especially with diamond, emerald, opal, and sapphire. He gives a good account of American finds of diamond, and a long account of European regalia. The book is full of interesting comment and contains many references to older authors.

The Tears of the Heliades or *Amber as a Gem*, by W. Arnold Buffum, G. P. Putnam's Sons, N. Y., 1900, is as its name implies a monograph on amber.

A good work on the history of precious stones and on historical jewels is *Gems and Jewels* by Madame de Barrera; Richard Bentley, London, 1860. It deals also with the geography of gem sources. An interesting chapter on "Great Jewel Robberies" is also included.

Of still greater age but of great interest is John Mawe's old work on diamonds and precious stones. In it the author discusses in a conversational style that is very attractive much of the gem lore of his day and shows a profound knowledge of his subject, a knowledge that was evidently first hand and practical, *A Treatise on Diamonds and Precious Stones*, by John Mawe, London. 2nd edition. Printed for and sold by the author.

For readers of French, Jean Baptiste Tavernier's *Voyages*, in six volumes, will be vastly interesting. Tavernier made six journeys to India and the East between 1640 and 1680 as a gem merchant during which time he purchased and brought back to Europe many celebrated gems including the famous French blue diamond which he sold to Louis XIV and which was stolen at the robbery of the Garde Meuble during the French Revolution. Tavernier describes these famous stones and many others that he was privileged to inspect in the treasuries of the Grand Mogul. He also describes interestingly and at great length the curious manners and customs of the people of the East. *Les Six Voyages de Jean Baptiste Tavernier*, etc., Nouvelle edition, Rouen, 1724.

Pliny's *Natural History*, to go much further back, is full of references to gems, and gem students should run through it (it is to be had in English translation) for such interesting bits as that in which he describes the belief that quartz crystal results from the effect of very great cold upon ice, a belief which Pliny himself is careful not to subscribe to. He contents himself with relating what others believe in this regard.

Both the Hebrew scriptures and the New Testament afford many references to gems with which the eager student of the subject should be familiar. "She is more precious than rubies" (referring to wisdom) is but one of these.

In conclusion the author hopes that this little text may lead a few to pursue further this most fascinating theme and that the pursuit may bring much of pleasure as well as of profit.

INDEX

A Book of Precious Stones

The Identification of Gems and Gem Minerals and an Account of Their Scientific, Commercial, Artistic, and Historical Aspects

By Julius Wodiska

8vo. With 32 Full-page Illustrations and 4 Colored Plates

A description, in altogether a new fashion, of gems and gem minerals, their nature and history, comprehensible to every reader, and of prime value to students and to jewelers.

The general reader will enjoy the simple descriptions of the origin, development, and treatment of the diamond, sapphire, and other precious stones, as well as of the beautiful semi-precious stones. Just enough of the technical has been provided to make the new gem book a *vade mecum* for students of gem minerals and for the army of jewelers in the United States, as well as their fellow-craftsmen and merchants in all English-speaking places. The art and industry of mounting gems is somewhat elaborately covered, especially as exemplified in the work of students at technical schools and the many unattached workers in jewelry designing and making who form a part of the Arts and Crafts movement. Some of the quaint superstitions about gems in the chapter on folklore have a curious interest. The author takes cognizance of the public desire nowadays for the novel and uncommon in gems, and shows that prospectors, gem miners, mineralogists, and jewelers are co-operating to greatly lengthen the lists of popular semi-precious stones. A chapter is devoted to collections of gems in museums.

Spotted something that isn't quite right?

Let us know and we'll fix it for future reprints. And as a thank you for your diligence, we'll send you a free Sothis book of your choice.

publisher@sothis.press

This page intentionally left blank

Except of course for that line above. And this one.
And whilst we have your attention, perhaps you'd
like to follow us on Instagram @sothis.press. Go on.

Made in United States
Orlando, FL
05 November 2026

90296313R00199